CLUBBING

Clubbing: Dancing, Ecstasy and Vitality explores the cultures and spaces of clubbing. Divided into three sections – 'The beginnings', 'The night out' and 'Reflections' – *Clubbing* includes first-hand accounts of clubbing experiences, framed within the relevant research and a review of clubbing in late 1990s Britain.

Malbon focuses particularly on the unwritten codes of social interaction among clubbers, the powerful effects of music and the role of ecstasy, clubbing as a playful act but also as a form of resistance or vitality, and personal interpretations of clubbing experiences. Offering an informative and intimate insight into the world of clubbing and the experiences of clubbers, this book presents a clear academic framework for study in this field and will also be relevant to those interested in popular and contemporary youth cultures more generally.

Ben Malbon is an Account Planner in an advertising agency. Previously, he completed his Ph.D. on clubbing at University College London.

CRITICAL GEOGRAPHIES

Edited by

Tracey Skelton, *Lecturer in International Studies, Nottingham Trent University*

Gill Valentine, *Senior Lecturer in Geography, The University of Sheffield*

This series offers cutting-edge research organised into three themes: concepts, scale and transformations. It is aimed at upper-level undergraduates and research students, and will facilitate inter-disciplinary engagement between geography and other social sciences. It provides a forum for the innovative and vibrant debates which span the broad spectrum of this discipline.

CLUBBING

Dancing, ecstasy and vitality

Ben Malbon

London and New York

First published 1999
by Routledge
2 Park Square, Milton Park, Abingdon, Oxon, OX14 4RN

Simultaneously published in the USA and Canada
by Routledge
270 Madison Ave, New York NY 10016

Routledge is an imprint of the Taylor & Francis Group

Transferred to Digital Printing 2005

Typeset in Perpetua by Routledge

British Library Cataloguing in Publication Data
A catalogue record for this book is available from the British Library

Library of Congress Cataloging in Publication Data
Malbon, Ben, 1972–
Clubbing: Dancing, ecstasy and vitality / Ben Malbon.
(Critical Geographies)
Includes bibliographical references and index.
1. Young adults – Great Britain – social life and customs.
2. Discotheques – social aspects – Great Britain.
I. Title. II. Series.
HQ799.8.G7M35   1999   99-12888
305.242'0941–dc21   CIP

ISBN 0–415–20213–2 (hbk)
ISBN 0–415–20214–0 (pbk)

Printed and bound by Antony Rowe Ltd, Eastbourne

FOR CHRIS, IAN, JON, JUSTIN, NICK,
RICHARD, RUSS AND THE TIMS.
THANKS.

CONTENTS

CONTENTS

PREFACE

He began by being an observer of life, and only later set himself
the task of acquiring the means of expressing it.

(Charles Baudelaire, 1964)

The Tunnel Club[1], Saturday night, summer 1997[2]

8 p.m. – Oxford Circus tube station, West End, London

It was balmy and still light, and very very busy. There he was. Seb, wearing velvet
trousers and a very bright velvet shirt, as he had promised. We shook hands. I was
excited, he was excited, but I could tell that he was tense about the evening; I'm
certain he could tell the same about me. He said I looked good and I was relieved
as I'd not known quite what to wear. He mentioned how odd it was to be going
out clubbing with me, a complete stranger, and I said the feeling was mutual and
that kind of let us both relax a bit. This was in some ways a completely artificial
situation – we were both strangers to each other – but I had good vibes anyway.
We went round the corner into Kingley Street to Brasserie Breton and grabbed a
few beers, chatting, taking up where we'd left off ten days previously when we'd
met in a bar at Euston. The whole thing was a little nervous to begin with – Seb's
a great story teller and I suppose nerves and beer combined mean that once you
start talking you just carry on.

On our way to the club…I was beginning to feel comfortable, like I was with a
regular clubbing buddy. Seb turned to me and said (and I still remember this
vividly), 'Ben, the difference between this and an ordinary night is that if you
don't like it, it's tough! You'll have to rough it out, whereas if I was with a clubber
mate, then I'd leave if he wasn't enjoying it. But tonight is MY night!' Back to
work…

11 p.m. – The Tunnel Club, queuing outside

What a feeling. Excitement, expectation, apprehension, tension. We started queuing at 10.30 p.m. and were about fifty people back from the front. By the time 11.00 p.m. came, we were relatively near the front of a large queue of about 200 clubbers. There was a definite nervousness or edginess to the atmosphere. We met Seb's friends in the queue and they were cool and really friendly. Suddenly the 'club host' came out and shouted something like this:

> Okay kids, as you know, we operate a really strict door policy at The Tunnel. We know the kind of party people that we want – the guys are all Elvis and the girls are all Marilyn. So I don't want you to get too down if we don't let you in – we've got a lot of experience of knowing what party people are like. If, when you get to the front of the queue, we ask you to sing a little Elvis song or something like that, then show us what a party person you really are, okay? Great.

As usual, I was getting a little worried that I wouldn't get in at this stage. Meanwhile, two six-foot transvestites were walking up and down the queue – occasionally one of them would ride a kid's chopper bike – talking and flirting with the clubbers. To be honest, I think more people were worried about actually getting in than whether the people inside were 'party people' – the whole thing certainly seemed designed to create an aura of exclusivity. Anyway, we got to the front and Seb started chatting and joking with the host guy. Needless to say, we got in without difficulty after that – I'd have to save my version of 'In the Ghetto' for another day.

Midnight – in the club

Seb said that he wanted to show me as normal a night as possible, although I knew this would not be completely possible. So, anyway, we set off to meet the staff, which was always the first thing that he did when he arrived at clubs. I felt like Ray Liotta's girlfriend in *Goodfellas* (well, kind of anyway). Everyone we met, it was: 'I'd like you to meet my friend Ben', 'This is a good friend of mine, Ben' and so on. Everyone was friendly and seemed to know Seb. We went to the loo and there was a guy in there specifically stationed next to the sink to put soap on your hands for you and wipe them afterwards. There was a selection of dozens of after-shaves and face creams that you could use if you felt the need. I'd always thought of this as the height of tack, but Seb actually knew the guy so that changed the way I looked at him – he was just a regular guy. These introductions were a really excellent thing for Seb to do – everyone seemed to remember him – and I felt both safe and lucky to be out with him. Really lucky. And yes, like everyone else who was there, I was getting fairly excited about the coming night.

Inside, the club was quite slick – a 'proper' club as opposed to the dingy cellars that are often the venues for London club nights. Three floors: a basement playing funk, hip hop and kind of easy stuff. Also a cloakroom, loos and a sweet shop down there. The middle floor had a grand foyer and huge staircase, permanently covered with people sitting, chatting, smoking, looking, moving. There was a big dance floor with a balcony all around (effectively the third floor). The balcony had big leather sofas and another bar. The DJ booth was behind a kind of elevated altar-like mixing desk and the music was full-on house, with the odd bit of garage. The whole club was dark, very dimly lit, though there were plenty of UV lights. No strobes. Tiny bit of smoke, not enough to have any real effect. About 500 people, 90 per cent white, mostly aged between 18–26 I guess. The clubbing crowd seemed relatively uniform – to the extent that many of the clubbers actually seemed to be *wearing* a uniform. Everyone was very dressed up: nearly all the guys had shirts and trousers on (jeans and trainers were not allowed at all). Most of the women wore very little – short skirts, vests, bikini tops, tans, gurning grins…

Later that night

Seb and his friends had taken their first E at about 11.30 p.m. and by 12.30 p.m. a few of them had taken their second. They weren't happy. The pills appeared to be duds. The placebo effect had lifted them all for a while, but they were no beginners and soon knew that nothing was happening. Seb rushed off to get some more pills and came back about thirty seconds later – no problem, sorted. By 12.30 p.m. the club was full and the music was getting progressively louder.

I tried to leave Seb and his friends to 'come up' on their pills alone so they didn't feel too much like lab rats under observation. I wandered. I suppose of all the areas within the club, the music downstairs was most to my liking. The highlight of my hour or so alone down there was a long and very loud version of Massive Attack's 'Unfinished Sympathy', which always sounds sensational on a large PA. I felt, fleetingly, at home, despite being alone in a new club. Dancing to music I loved helped me forget about the novelty aspect. Without music, dancing is nothing. It's not really possible. In this respect the music *is* the experience, yet clubbing isn't just about listening, it's very much about *doing*. Is music then purely bodily? I had another beer, then waded into the crowd again.

2 a.m. – main dance floor, chaos...

The music dominated the dance floor. Everyone was dancing – on the balcony, on the little stages that projected out onto the dance floor like catwalks (look at me, 'cos I'm looking at you!), in the bar, behind the bar, *on* the bar. I really enjoyed dancing. I felt myself slipping in and out of submission to the music. No sooner had I forgotten what I was doing, and my dancing had become almost automatic, than I was suddenly aware of myself again, conscious of moving my feet, looking at what my arms were doing. I looked at people dancing and noticed how overtly they were looking at everyone else. I don't just mean glancing either. I mean really *looking* at someone, as though that was completely normal. I could feel myself being scanned, but wasn't affronted or anything by this. We all seemed to want the music to take us over; to *become us* in some way. Okay, we each stamped our individuality on it in our own way – a neat little step here, an arm movement there – but the clubbers were essentially doing the same thing as each other and in the same place and at the same time. Certain tracks that were clearly club favourites were greeted rapturously by the clubbers, and I found it easy to become enthused. The atmosphere was contagious. No sense of embarrassment or coyness – every man was Tony Manero from *Saturday Night Fever*, every woman was Lydia from *Fame*. Sweat becomes something to be proud of rather than a social stigma.

4 a.m. – lost for words, lost in time and space, just lost...

Wow. There came a point when I was just taken aback by what I was witnessing, of what I was actually partly constitutive, of what *we* were and had become. Some form of extraordinary empathy was at work in that crowd, particularly when at the kind of extended climax of the evening the music and lighting effects combined so powerfully with the moving crowd on the dance floor. Clubbers were losing it all over the place. This kind of context – this sound and lightscape – must surely significantly change the ways that people interact. I mean, people are just so close to each other; proximately and emotionally. The clubbers were sharing something precious, something personal, something enriching. The intensity of this fusion of motions and emotions was almost overwhelming.

How can words – simple, linear words on a page – evoke this delirious maelstrom of movement and elation? Again and again I arrive at this point in 'The night out' and I simply cannot describe it any further. How can I convey the deep, thundering bass which is felt more than heard? The mass of bobbing bodies: blurred, colourful, dimly-outlined and unceasingly in motion? The space itself, which fleetingly seems as though it has no edges, no end in time or space, yet at the same time only stretches as far as you can see into the lights, the black walls, the heaving dancing masses? The sensation of dancing, of moving without thought, of moving before thought, of just letting go, letting it all out? Words fail me; words become redundant and unnecessary, words become pointless.

4.30 a.m. – back into the city

Suddenly, it's over. Exhausted, Seb and his friends were leaving to go and calm down a bit somewhere, drink Absolut and smoke a few spliffs. They accosted a cab outside and he tried to persuade me to go with them, but I was completely out on my feet and didn't want to intrude on their chill-out time so I stumbled towards Trafalgar Square which was, as usual, completely heaving. I was certain that there were more people in the Square than there are at four in the afternoon. My ears were ringing horrendously – I felt like a television that had been left on after close down. Seduced by the aroma of cheap sausages at a hotdog stall, I stuffed one down in three mouthfuls, before it had a chance to taste shit.

The night bus was…well, just the night bus. Overcrowded, full of mostly young people, half of whom were nattering like idiots and half of whom were virtually asleep. No one really looked what you'd call 'healthy'. I remember slumping back under the flickering, fluorescent lights of the bus, my sodden shirt suddenly clinging to my back. I felt bloody dirty. Dirty, tired and cold. Ears still ringing. Chest pains from smoking thirty fags. Hearing and sight affected.

5 a.m. – Vauxhall Bridge, South London

At Vauxhall Bridge I reluctantly heaved myself off the bus. After the bus had crawled slowly away, it was eerily silent, empty apart from a couple spilling out of the Dungeon club, laughing and messing about, talking smuttily about sex toys. At this time the night's really over, yet there's an hour or so until Sunday starts properly. I couldn't believe it was getting light already, although the street lamps still cast a faint orange glow on the pavement. The chocolate-brown Thames snaked silently beneath my feet, the pedimented grandeur of Pimlico behind me, the demented mess that is Vauxhall ahead. In-between days in in-between places.

An occasional black cab rumbled past. The planes had just started and London looked wonderful as it awoke. The sky was an intense and brilliant blue with streaks of yellow just beginning to show themselves behind the MI6 building as I passed it to my left. It was spookily quiet. I thought I could hear individually each of the four engines on the jet flying overhead, which I was convinced was flying far too slowly to stay aloft. I craned my neck in a kind of child-like gesture to gawp skywards. That was the only noise – a solitary jet in the sky and the diesel engines of the odd taxi. And my weary steps. I thought about Seb and wondered what he was thinking. I wondered when we'd meet again.

THANKS

This book could not have been written had it not been for the help of a group of clubbers who gave up both their time to talk to me and their nights out to share with me. I feel extraordinarily privileged to have met such a fun bunch of people; they energised this project, and me, with their enthusiasm for clubbing.

The Ph.D. research, on which much of the book is based, was carried out while I was at University College London. At UCL, Phil Crang was an unfailing source of encouragement, momentum and patience amidst my confusion. Kemal Ahson, Tracey Bedford, Kevin Collins, Nicole Dando, Luke Desforges, Claire Dwyer, Minelle Mahtani, Mark Maslin, Simon Pinnegar and Peter Wood have all provided stimulation and words of encouragement, and, more importantly, have been good friends. Peter Jackson, Simon Pinnegar and Mary Anna Wright offered comments, which were particularly incisive and helpful, on earlier drafts and thanks also goes to Jacquie Burgess, Miles Ogborn, Susan Smith, Nigel Thrift and Sarah Thornton, all of whom offered advice and provided yet more ideas at various times. Gill Valentine deserves a special mention as the first to show real interest in the project, and I thank her for her support over the years.

Back in the real world, and important for different reasons, huge thanks go to Carol Barrett, Bat, Martin Berger, Jude Hill, Jamie Johns, Tim Malbon, Nick Oram, James Petrie, Chris Quigley and Alexis Richardson. All have had otherwise excellent nights (and plenty of days) marred by my confused mumblings and one-track mind. Jon Malbon was there when it all began and may have been the wisest and coolest of us all. His early influence was crucial – I should have seen the signs when he used to keep me awake with his repeated playing of 'My House is your House' at 4 a.m. Katie has, as ever, been the best buddy one could ever hope for, and I love her to bits.

Not surprisingly, I have found many writers to be inspirational, but there are four that stand out in particular. The writings of Marghanita Laski completely blew me away when I first encountered them, especially her classic 1961 text, *Ecstasy: A Study of some Secular and Religious Experiences*. Michel Maffesoli's sociology, which focuses on the claimed resurgence in the vitality of crowd experiences in the late twentieth century, has also been inspirational. A third writer who has been highly influential is Erving Goffman. His massively incisive, yet also subtle and totally accessible studies of how and why we present ourselves to others in various social contexts continue to amaze me, and his *The Presentation of Self in Everyday Life* (1959) remains the most incredible book I've read. Finally, special thanks must go to Nicholas Saunders for his seminal work on understanding and demystifying Ecstasy, and for trying to provide clubbers with the information that they need and demand. Although, tragically, he died in a car accident during the research on which the book is based and although I never met him in person, much of the optimism that runs through the book is thanks to him.

Without the financial assistance of the Economic and Social Research Council (award number: R00429434210) and the University College London Graduate School, I would not have been able to complete the research on which the book is based. I want to acknowledge their vision in providing me with this opportunity. *The Face* and *i-D* magazines were the only publications to print my letters requesting clubbing volunteers and I am indebted to the editors of each for providing what to them was a tiny service, but what to me turned out to be a vital break. At Routledge, Sarah Carty has been both patient and enthusiastic.

My earliest musical memory is of my parents playing the soundtrack to *Saturday Night Fever*, although always after I had gone to bed (what is it about my family!?). Although they'd be the last to agree, my parents have been the coolest parents one could hope for and I thank them for their individuality, love and freedom.

Now, about that early night I promised myself...

London
January 1999

Part One

THE BEGINNINGS

To be away from home and yet to feel oneself everywhere at home; to see the world, to be at the centre of the world, and yet to remain hidden from the world – such are a few of the pleasures of those independent, passionate, impartial natures which the tongue can but clumsily define.

(Baudelaire, 1964)

THE NIGHT AHEAD

This book is about clubbing. More specifically, this book is about the experiences of going clubbing. It is concerned with some of the motivations for and the socio-spatio-temporal and bodily-emotional practices which constitute the clubbing experience. The book is also concerned with many of the cultures, spaces and mediations influencing and 'producing' the clubbing experience. Finally, this book is concerned with sketching out an understanding of the vitality that may be engendered through the experiences of clubbing.

The book sets out to answer three distinct, but closely related questions:

1   How is clubbing constituted through the practices, imaginations and emotions of the clubbers themselves?
2   How can music and dancing so powerfully affect our experiences of certain spaces, of ourselves and of others?
3   How is clubbing, as a form of 'play', significant within the identities and identifications of the clubbers, and in what ways can it engender vitality through its playful practices?

In attempting to answer these questions the book is split into three parts. Part One is comprised of a number of introductory sections which I have called 'The beginnings' – each section within this initial part of the book performs an important role in contextualising 'The night out' to come. So, after this short introduction to the book, the next section in 'The beginnings' provides a contextual background for the project in the form of a brief review of the state, scope and scale of clubbing in late 1990s Britain. I follow this contextual introduction by setting out in some detail the three major academic starting points for 'The night out'. These starting points are: young people at play, consumption and consuming, and the sociality and performativity which arise out of a concern with processes of identity formation and amendment.

After these 'Beginnings', the main body of the book, Part Two, comprises an increasingly complex schematic and thematic move through some of the multifarious times, spaces and practices of clubbing. This second part is presented along the broad lines of a night out, beginning with club entry and finishing at the end of the night. Each of the four sections of 'The night out' forms a cumulative spotlighting of what I understand to be the key thematic elements that comprise one approach to understanding the constitutive practices of clubbing. Each section builds on the last in introducing a further level of understanding or an added slice of complexity. I turn, in the first section of 'The night out', to issues around 'belongings' and notions of identification. Some of the reasoning behind clubbers' desires to go clubbing are discussed, and I develop this through a brief conceptualisation of 'coolness' and 'style' and an examination of the importance that negotiating entry to the club itself can play in the establishment of these belongings. In the second section of 'The night out', I progress this interest in belongings by looking in detail at the importance of music and dancing in constituting clubbing crowds. I draw out, for special attention, the clubbers' fluctuations between experiences of self and of the crowd, and the timings and spacings of their dancing practices. These clubbing crowds of intense motion and emotion can provide the context for moments of extraordinary euphoria and notions of freedom. Thus, in the third section of 'The night out', I address the experiences of what have often been referred to as 'altered states', but which I attempt to refine in describing as 'oceanic' and 'ecstatic' experiences. These experiences are explicitly attained partially though the use of drugs, such as ecstasy (MDMA), and I examine in some detail the use of ecstasy in the clubbing experience. I complete 'The night out' by building on these conceptions of alterity in the fourth section of 'The night out', where I introduce the central notion of the book – playful vitality. This is a conceptualisation of the sensation of inner strength and effervescence, which, I argue, can be experienced through the practices of 'play' and especially through the 'flow' achievable through dancing. Playful vitality is conceptualised as an alternative approach to understanding the nature of and relationships between notions of power and resistance.

The third, shortest and final part of the book is Part Three, 'Reflections'. In one respect the night out ends abruptly as the clubbers leave the club. Yet there are important post-clubbing processes of reflection and attempts at understanding which many clubbers go through, even if they are alone. In addition to tracing a variety of differing routes for the clubbers through these 'Reflections', I also reflect upon the night out that has just occurred. Re-visiting the three starting points which I have briefly set out in 'The beginnings', I draw out a number of key themes which emerge from returning to these starting points through the lens of 'The night out'. Clubbing offers a myriad of insights into our conceptualisations of a whole range of social interactions, notions of communality and play, and of being young. Far from being a mindless form of crass hedonism, as some commentators suggest, clubbing is for many both a source of extraordinary pleasure and a vital context for the development of personal and social identities. Yet, I argue, our understandings of clubbing, its practices and its remarkable resonance within the lives of so many young people, are only just beginning.

As far as reading the book is concerned, while 'The beginnings', which together constitute Part One, comprise a sketching of some of the boundaries of and context for the book in academic terms, they are not critical to an understanding of 'The night out'. That said, while these 'Beginnings' may therefore be skipped, they remain central in contextualising the particular understanding of clubbing that I present here.

CLUBBING CONTEXTS

Clubbing is a hugely significant social phenomenon, as anyone walking around the centre of a city or town in the UK at 2 a.m. on a Sunday will attest. Clubbing is notable primarily because of the sheer scale of its appeal, and the increasing eclecticism of its constituent genre. Yet clubbing is also notable because of its 'systematic demonisation' within the media and the introduction of new legislation inhibiting clubbing – the Criminal Justice and Public Order Act (1994) for example – and the widely-publicised, although somewhat over-hyped, involvement of illegal practices, such as drug use, in clubbing (Ward, 1997).

Clubbing is an overwhelmingly urban form of leisure and is now a major cultural industry (Lovatt, 1996)[1]. Few towns or cities are without at least a couple of clubs, with many cities such as Manchester, Leeds, Glasgow, Liverpool, Sheffield and London having well-developed and multi-layered clubbing industries which contribute substantially to the local city economies (O'Connor and Wynne, 1995). Indeed, many city authorities have actively pursued programmes of inner city regeneration premised partially on the economy of the night and the attraction of thousands of clubbers with relatively high disposable incomes into these areas.

Clubbing is now also an increasingly international leisure pastime, and it is not uncommon for clubbers living and clubbing in the UK to spend their annual holidays clubbing in another, usually warmer and often cheaper, part of the world. The explosion in clubbing cultures over the last ten years has thus been accompanied by – and undoubtedly further fuelled through – the ever-widening horizons of some of the clubbers themselves (Williamson, 1997). Williamson (1997) suggests that the 'top eight destinations for British clubbers' are: Ibiza (Majorca), Goa (India), Guadeloupe (Caribbean), Cape Town (South Africa), Tokyo (Japan), Ko Pha-Ngan (Thailand), Singapore (Asia) and Sydney (Australia). Many of these locations have historically been popular with the 'alternative' traveller, but in recent years they have increasingly been appropriated by clubbing holiday-makers. While these evolving cultures of clubbing 'away-from-home' would seem to support points that I make later in respect to alternative social orderings, senses of other-worldliness and escape, this text is restricted in scope to spotlighting the clubbing experiences of clubbers in central London.

The many facets of clubbing in London are changing dramatically, as they always have done (Kossoff, 1995; Thornton, 1995). On a typical Saturday night in 1986, a flick through the pages of the London listings magazine, *Time Out*, might have revealed a choice of twenty clubs (Collin, 1996b). Thirteen years later this choice has grown to in excess of fifty clubs and this total excludes 'student' and 'gay' clubs, each of which now have their own sections within the listings. Also excluded from this total are late-night bars with dance floors, the numbers of which are growing rapidly and even starting to pose a threat to clubs, unlisted nights of which there are surprisingly many and clubs much beyond London Transport travel zone 3 (about 6–7 miles from central London), beyond which *Time Out* takes little interest. With all these taken into account, there may be anything between 400 and 500 club nights on offer every week in Greater London, with the highest number available on Bank holidays when Sunday effectively becomes an additional Saturday – the busiest night of the week.

There are no London-specific studies covering the growth of and recent developments within clubbing as an industry. However, despite tending towards generalisations at times, recent marketing intelligence studies by the market research company, Mintel, provide a valuable and large-scale overview of the 'nightclub and discotheque industry' – hereafter the 'clubbing industry' – in the UK (Mintel, 1994; Mintel, 1996)[2]. In particular, these studies provide detailed figures concerning both the continued explosive growth and also the changing structure of the clubbing industry in the UK. However, it should be emphasized that Mintel do not distinguish between nightclubs and discotheques when calculating market size and commenting on the industry as a whole. They define both nightclubs and discotheques as 'establishments which offer music, drinks, food [where there is a legal requirement to do so], dancing and lounging under one roof' (Mintel, 1996: 1). Thus, Mintel's analyses encompass nightclubs and discotheques of all sizes and forms, from huge chains of franchised clubs run by companies such as Rank Leisure, Scottish & Courage PLC and Granada to the tiny, independent and single-site clubs that make up the majority of clubbing establishments in the UK. The facts and figures provided by Mintel – some of which I refer to later – are thus concerned with the nightclubbing industry as a whole, whereas my interest in the 'The night out', which forms Part Two, is much more concerned with the *Time Out*-style clubs (see p. 35) located in central London. That said, the statistics and trends provided by Mintel are indicative of the very large scale of the clubbing industry as a whole and are thus useful in sketching this broad overview of the nature of clubbing. Neatly paralleling and complementing the Mintel studies, a recent Release survey (1997) interviewed

520 clubbers for a study of the relationships between clubbing and the use of drugs[3]. I draw from both these sources in the discussion that follows, as well as throughout the book.

In crudely financial terms, revenue generated by the clubbing industry was forecast to have broken through the £2 billion per annum barrier for the first time in 1996, a 6 per cent increase over 1991 revenue levels, although still 14 per cent less in real terms due to the impact of the recession in the early 1990s (Mintel, 1996). A total of 42 per cent of the general population of the UK now visit clubs (or discotheques) at least once a year, compared with 34 per cent in 1991, and 43 per cent of 15–24 year-olds visit a club once a month or more often (Mintel, 1996: 5). The trend over recent years has been for more people to visit clubs, but less frequently, with the average spend per visit being £11.60 per head, which is still considerably lower than the 1991 average of £13.77 (Mintel, 1994; Mintel, 1996). This reduction in average spending is partly explained by a growth in the significance of mid-week clubbing – Monday through to Thursday inclusive – which attracts lower spending per head because of discounted alcohol and admission charges, as well as higher numbers of students. Late 1998 and early 1999 saw the opening of a number of very large club-bar-restaurants (or super-bars) in central London, which are all financed by large entertainment and brewery-based companies[4]. Increasingly, it appears that clubbers are demanding more from a night out, and are no longer content with being charged £10 or £15 to be allowed to dance. It remains to be seen whether the popularity of clubbing as it currently stands will endure these new developments in night-time entertainment.

Structurally, Mintel suggests that the clubbing industry has been badly affected by the early 1990s recession, with the trend towards weekday admissions and discounted admission prices partly reflecting this. A reduction in the total number of clubs from 4,200 in 1994 to 4,100 in 1996 not only reflects the impact of the recession, but also – given the overall increase in admissions over this period – the general trend towards larger club capacities, and an increasing concentration of clubs within the portfolios of large leisure organisations (Mintel, 1996).

It is not immediately apparent that many clubs are owned and run by large national chains, for example Rank Leisure and Granada, because few clubs share nationally-known brand names, usually very little is made of the network of which a club might be just one venue, and accordingly marketing is usually carried out on a local basis (Mintel, 1996; Thornton, 1995). However, increasing numbers of club operators are turning to direct mail in order to target clubbers more aggressively (Mintel, 1996).

Release found that most (85 per cent) of those sampled at what they called 'dance events' were under 30 years old, and fell into the 20–24 age-range, a figure which supports the Mintel survey results. Mintel (1996) suggests that a growing number of 25–34 year-olds are clubbers, although the frequency of club-bing visits and the total number of clubbers falls away dramatically with increasing age, with the mid-20s appearing to be a watershed.

Most (78 per cent) of Release's 520 clubbers started clubbing during their teenage years, with just 8 per cent starting to go clubbing *after* the age of 24. As far as motivations were concerned, while music (45 per cent), socialising (37 per cent), the atmosphere (35 per cent) and dancing (27 per cent) were the top four in the Release survey, these were very closely followed by drug use (22 per cent). Release suggests it is interesting to note that only 6 per cent of clubber respon-dents claim 'meeting prospective sexual partners' to be important. Men were more likely to say that they enjoyed clubbing for the drug use, while women were more likely than men to suggest 'socialising' as important. Socialising also appeared to be more important for the older clubber. While 28 per cent of 15–19 year olds mentioned socialising, this rose to 45 per cent for those aged over 30 years (*Release Drugs and Dance Survey: An Insight into the Culture*, 1997). In contrast to Release, however, Mintel proposes that single consumers were six times more likely to go clubbing than those who were married, with 19 per cent of single people agreeing that clubs were a good place to meet a prospective boyfriend or girlfriend. This divergence between Mintel and Release over the significance of clubbing as a forum for meeting sexual partners can be partially explained by the breadth of the Mintel survey. The Release survey was restricted to what might be called 'dance clubs' – as is this book. However, the Mintel survey, by its very nature, encompassed all types of nightclub and discotheque establishments, from the 'dance clubs' of Release, where drug use appears to result in what might be labelled 'a slight muffling of the libido', to the Ritzy discotheques that constitute a very different form of clubbing experience[5].

As far as generalisations about social class and geographical variation are concerned – and are worth – Mintel (1994; 1996) suggests that frequent visitors to clubs are drawn predominantly from the 'middle income C and D socio-economic groups', and that those living in the northern regions of the UK, especially within the large northern cities, go clubbing, on average, slightly more often than their southern counterparts.

In terms of clubbing genre, 'gay' clubbing is marketed to a quite specific frac-tion of the clubbing population and receives a separate listing space in most listings magazines. However, it should be stressed that many non-gay clubbers enjoy the atmosphere they find at gay clubs and, once again, this is a point that I discuss later. After a broad division along the lines of sexuality, clubs are most often distinguished by both clubbers and listings sources along the lines of musical types and night of the week. A comprehensive survey of the changing and contin-ually fracturing musical genre that constitutes clubbing is beyond the scope of this project. In any case, given the continuing explosion in genre, sub-genre and sub-sub-genre, a survey of this nature would be obsolete within days. However, it should be noted that musical genres continue to fragment, and, despite the continuing prevalence of clubs playing techno, house and jungle (or drum'n'bass), none of these are, at the time of writing (late 1998), particularly in the ascendancy. This is a notable development from, say, even five years ago when most clubs played either house or techno music, with only a minority playing other musical forms. As Collin says:

> [O]nce there was simply house. Now genres like jungle, techno, ambient and trance have developed so far from their origins that they are barely traceable to their source…there is no overall blueprint, no masterplan, just an endlessly shifting set of possibilities, fracturing scenes and fracturing ideologies.
>
> (Collin, 1996b: 112)

Clubs are currently far more likely to offer an eclectic blend of music within a single club night than just one choice. Some rooms may be set aside to play house or techno music, while others play jazz or drum'n'bass, or perhaps more soothing 'loungecore' or easy listening. In short, 'genre-lisations' about the musical trends within the cultures of clubbing are impossible to make – the scenery changes too quickly. What is certain is this: that only twenty years after disco's popularity went into orbit in the wake of *Saturday Night Fever* (1977) and only ten years after the 'acid house' revolution of summer 1988, clubbing is now an immensely big business, remains extraordinarily popular, and is increasingly fragmented in form.

THREE STARTING POINTS

Did ye not hear it? – No; 'twas but the wind,
Or the car rattling o'er the stony street;
On with the dance! let joy be unconfined;
No sleep till morn, when Youth and Pleasure meet
To chase the glowing hours with flying feet.
(Byron, 1816)

The approach to developing an understanding of the practices and spaces of clubbing that is presented in this book is grounded within, and in many ways
develops, certain existing areas of debate and research on how and why young
people spend their free time as they do. Broadly speaking, there are three main
points of departure for this book in academic terms – three thematic 'starting
points' which bubble up throughout the book, yet never quite break the surface
explicitly in 'The night out', which forms Part Two of the book. In one sense,
these concerns are obvious, for, like clubbing itself, they are what the book is
'about', they *are* the book.

The approach which I present in this book is related to a set of debates, literatures, academic disciplines and sub-disciplines that are concerned with the
changing geographies of young people's leisure and lifestyles, as well as the
consuming activities through which these lifestyles are constituted – materially,
but crucially also experientially and imaginatively. A central theme of 'The night
out', which forms the main part of the book, is a concern with this constitution
of clubbing as simultaneously practical and emotional.

The first two starting points are geographies of young people at play and
broader, but related, geographies of consuming. I am especially interested in the
relationships between processes and practices of consuming and notions of
identity and identification formation and amendment in the clubbing experience. Therefore, a third starting point, which spins off these first two, is

11

geographies of sociality and performativity. I now address each of these in greater depth.

Starting point one

Geographies of young people at play and of clubbing

To understand the behavior
of someone who is a member of...a
group it is necessary to understand
that way of life.

(Becker, 1983: 79)

The man bought her Campari and, fascinated, she watched the ice lumps in her glass leap into brilliance, their veins outlined in dazzling blue, then die. The music went bang-bang-bang-bang; it filled her cranium, as if the whole city was gathering itself up into a concerted heart and thumping in her head, bang-bang-bang-bang...'I like', she said.

(Raban, 1974: 239)

Young people and especially their languages, practices, spaces and cultures are poorly understood by the social sciences and to a lesser extent by the media. This is particularly so as far as young people at play are concerned, given that play is frequently overlooked as an irrelevant aspect of people's social worlds, however old the people in question might be. Through this first starting point I briefly address the birth of 'youth' as a category of academic study and interest[6]. I note how the growing popularity of studying 'youth' was closely associated with the rapid expansion of a mainly urban popular culture in the post-1945 period, and I briefly summarise some of the key developments in understandings of young people up to the present. Second, I turn explicitly to the research literatures on dance cultures and clubbing. I highlight both the recent profusion in work on clubbing, but also the generally modest impact of this work in aiding our understandings of the practices and spatialities, or geographies, of clubbing. Third, in the light of this brief survey of clubbing literatures, I highlight a number of key deficiencies in our understandings of young people as clubbers. These deficiencies have evolved in parallel with the growth in popularity of this form of consuming. In particular, I look at the notion of 'resistance', suggesting that it must be revisited and revised in the light of our changing understandings of young people and their contemporary social worlds.

Young people and the category of 'youth'

There have always been young people, but there has not always been a time in people's early lives known as their 'youth' – a descriptive term that denotes a period of 'in-betweeness' for young people. Part-child, part-adult, but falling neatly into neither group in society, youth as a category in its sociological form emerges in the 1920s, although Aries (1962) traces what he calls this quarantine period in the lives of the young back to the early eighteenth century (Skelton and Valentine, 1998). The emergence of 'youth' as a category of academic interest is largely credited as the work of the American tradition of ethnographic research and especially the work of the Chicago School on deviancy (see below) (Hebdige, 1983). Before these early sociological studies on the cultures of youth, young people were taken to be literally that – just younger versions of people who were in the process of becoming adults. Once people had 'come of age', they were regarded as full adults in society. The period of 'youth' was thus one of 'maturation' during which a young person would learn how to be young, how to behave as a person, and as an adult with responsibilities (Boethius, 1995).

The Sociology Department that was set up at the University of Chicago around the turn of the century based its investigations on qualitative empirical research (Gelder and Thornton, 1997). The micro-sociological interest of the school and an interest in the city more generally spurred members of the department into studying taxi dancehalls, waitresses, hobos, juvenile delinquents and thieves. Social interaction was always at the very centre of concerns, though very much social interaction in the context of what were seen as the social *problems* of the day. The city was thus given special prominence in the changing and complex social structures of society. The present study is not the place for an in-depth history of the rise, stigmatisation and more recent collapse of the category of youth; my interest is more in the direction of the emergent geographies of contemporary young people in respect of their clubbing practices. However, three interrelated aspects of the emergent sociology of youth, of which the Chicago School formed an early part, persist to the present and resonate through constructions of contemporary clubbing – youth as predominantly urban, as somehow deviant, and as in, or the cause of, trouble. The explosion in opportunities to consume, in consumption choices and especially in cultural influences from the United States augmented a general post-war optimism and fuelled an unprecedented rise in young people's interests in style and the practices of distinction (see p. 51) (Hebdige, 1988). Yet, the flipside of being young – youth-as-trouble – was never far below this stylistic surface, even if it was often momentarily obscured in the hype and hyperbole[7].

These twin notions – of youth as stylish and as somehow deviant – constituted the main concerns of those at the Centre for Contemporary Cultural Studies

(CCCS) during the 1970s and early 1980s. Based at the University of Birmingham and established in 1964, the CCCS also played a critical role in the establishment of 'youth', young people and teenagers as valid objects of academic study in Britain. Despite significant qualification and subsequent refinements of the theoretical tenets that emerged from the Centre, in many cases by the original Centre members themselves, many of whom have since moved on to new areas of interest, the work carried out by members of the CCCS continues to endure in numerous aspects of both theoretical and methodological approaches to young people[8].

For the members of the CCCS in the 1970s 'youth' became the central interest, and particularly the *styles* of youth, which were seen as somehow signifying class and 'subculture'. Aspects of British Marxist theory – the work of Thompson, Hoggart and Williams – were blended with continental influences in social theory – Gramsci, Althusser, Barthes – in creating a multi-disciplinary structuralist approach in which class and culture were central strands (Gelder, 1997). Notions of 'subcultures' and the 'spectacular', as well as the relationships between ideology and form – reified in 'styles' – were spliced with attempts to locate the subcultures in relationship to the broader cultural categories of the 'parent culture' (in the case of the CCCS, this was inevitably the working class) and the 'dominant culture' (mainstream youth more generally). Youth was constructed as a period of transition between the parent culture within the working class and a mass culture of increasing opportunities in leisure and entertainment. This mass culture was seen as a world of commercially driven concerns where one's class roots seemed progressively less important (Gelder, 1997). Furthermore, much early CCCS research was heavily focused on young men, with young women either ignored altogether or seen merely as peripheral (McRobbie, 1994b; Skelton and Valentine, 1998). By far the most significant methodological approaches to these studies of young people were those premised on semiotics and, to a lesser extent, ethnographies (see Clarke et al., 1993; Willis, 1977).

However, as the 1970s merged into the 1980s the confetti of critique was beginning to impact on both popular and academic understandings of young people, including those produced by the CCCS. In particular, Dick Hebdige's (1979) book, *Subculture: The Meaning of Style*, was the first text about young people and their styles really to exploit the academia/pop critic cross-over. Hebdige's book received a wide audience, stimulating debate around youth, style and music on an unprecedented scale. Questions were posed that had never been asked before, even during the heyday of subcultural theory in the late 1970s. For instance, who was doing what? Where did this 'style' come from? Where was it purchased, and who was selling it to whom? More abstractly, what were the social relations which informed the production of the subculture (McRobbie, 1994b)?

By the early 1980s, theorists were suggesting that the way forward for subcultural analysis lay in a revival of interest in the very *ordinary* nature of most subcultural lives and the practices involved in their constitution. In the introduction to the second edition of an earlier text, Cohen (1980) savages many of the approaches of the early CCCS, but he picks out the emphasis on spectacularity at the expense of the everyday for extra attention. Cohen proposes that this emphasis obscures occasions where styles might be conservative or conformative. In cases where there *did* seem to be a conservative impetus to the stylistic elements on display, the notion of *bricolage* was all too often used merely to re-read the apparently conservative meanings as really hiding just the opposite (Cohen, 1980).

Furthermore, uniformity of subcultural adherence was attacked, and particularly the notion that the subcultures somehow evolved spontaneously without any input from the media or from other cultural traditions, although Hebdige exempts himself from this charge (Cohen, 1980; Thornton, 1994). What was beginning to emerge as an obvious disjuncture between the closely homological theses of the CCCS theorists and the much messier and more complex sets of meanings and values of the members of the subcultures themselves was put down to an over-emphasis upon semiotics and certain strands of structuralism (Cohen, 1980).

By the mid to late 1980s, a number of individuals researching the cultures of young people – both at work or school, and at play – had revised earlier conceptualisations and set off in new directions. Most of all, these revisions had been prompted by the continued dramatic and far-reaching developments in the social lives of the young over the previous 10–15 years. McRobbie (1994b) suggests that the turning point represented by punk, and the shock waves that spread throughout the youth cultural worlds of consuming thereafter, especially those of fashion and music, meant that never again could youth cultures be simply seen to occupy a 'folk devil' status in society. Furthermore, a profusion in the total numbers of young people affiliating themselves with, or being ascribed membership of, youth subcultures has occurred. It is now increasingly the case that it is difficult to identify any young people who are *not* affiliated to some stylistic grouping or other. This has occurred in conjunction (and in synergy) with a continuing growth in media, film and technological influences upon young people; the latter particularly in the realms of video and music production/consumption, and even more recently the IT/Internet/multi-media communications interface. The rapid expansion in print media for the youth market has also been incredible, even spectacular. Young people and their styles now have their own dedicated industries with the growth and consolidation of a huge and still expanding style media. 'Club cultures' (Redhead, 1997; Thornton, 1995) or 'dance cultures' provide a particularly apposite example of these social–theoretical dynamics.

From subcultures to club cultures – understandings of clubbing

The latest and by a long way probably the largest and most influential of recent young people's cultures or styles in Britain can be found in club cultures (Thornton, 1995). British club cultures have their roots in a blend of international musical and cultural influences that stretch back, at least in easily recognisable form, to 1945 (Wright, 1998); the gay and 'black' disco scenes of New York in the mid to late 1970s were particularly important[9]. These dance cultures are now highly fractured in constitution and continue both to fragment and expand extremely rapidly. This growth is both in terms of sheer numbers of participants, but also in terms of the number of genres and sub-genres that are affiliated to them, and the wide variation in practices of sociality – the ground rules and customs – which each of these in turn demands and continually evolves (Mintel, 1996; Wright, 1998).

Despite their relative invisibility, at least compared with certain previous so-called 'subcultures', such as mods or punks, club cultures and clubbing – also called 'raving', 'dance culture' and 'nightclubbing' depending upon the age and experiences of the people you ask – are the focus of increasing numbers of research projects and texts. Commentators on 'youth' and young people are constantly reminding us of the central significance of this emerging social phenomenon. Thus, we can read that 'dance culture is...in opposition to dominant political and cultural forces' (Wright, 1998: 3); that 'these changes in youth culture are at the cutting edge of "politics" and deviance' (Redhead, 1993: 5); and 'rave seems to overturn many of the expectations and assumptions we might now have about youth subcultures' (McRobbie, 1994b: 168). However, many of these accounts are narrow in focus, with few going beyond either merely stating that clubbing is significant or – and much worse – simply indulging in painfully uninformative accounts of clubbing as 'the dark side of critical theory's dystopic moon' (Hemment, 1997: 24), as returning the raver 'to a pre-Oedipal stage, where libidinous pleasure is not centred in the genitals, but where sexuality is polymorphous and where sensuality engages the entire body' (Rietveld, 1993: 54) or – and most agonisingly naive – where 'in the loving space of the rave, young people are creating a potential blueprint for the whole of society to follow' (Richard and Kruger, 1998: 173). Crucially, as Gibson and Zagora (1997: 2) note with an admirably blunt brevity, 'as with many other youth musical subcultures, the voices and concerns of ravers are rarely heard in academic texts'.

A number of important contributions to evolving understandings of clubbing do stand out. In her detailed discussion of the cultures and differentiations within clubbing, Sarah Thornton (1994; 1995) builds upon the work of Pierre Bourdieu (1984) in presenting a discussion of 'subcultural capital' and the distinctions through which this capital operates in the cultures of clubbing. Using and extending Bourdieu's analysis of 'distinction', Thornton represents club cultures as 'taste cultures' (ibid.: 3). She draws a distinction between notions of the 'main-stream', and what she calls the world of 'hip' ('cool'), suggesting that the mainstream is 'the entity against which the *majority* of clubbers define themselves' (1995: 5; emphasis in original). Thornton then proceeds to uncover hierarchies working within these taste cultures through in-depth ethnographic analysis of the birth of 'acid house' music between the years 1988–92. This is complemented by a historical account of the 'authenticities' of records and recorded events since 1945. Thornton's contribution is a valuable attempt to rectify the treatment of young people and their cultures as being isolated from other 'outside' influences (McRobbie, 1991). As Thornton proposes:

> [C]ontrary to youth discourses...subcultures do not germinate from a seed and grow by force of their own energy into mysterious movements to be belatedly digested by *the* media. Rather, media are there and effec-tive from the start. They are integral to the processes by which, in Bourdieu's terms, 'we create groups with words'.
>
> (Thornton, 1994: 176; emphasis in original)

However, despite moving on to address the messy and shifting notions of 'cool-ness' through which young people position themselves, to a greater or lesser extent, in consuming the clubbing (or raving) experience, Thornton nevertheless largely neglects the *experience of clubbing* itself. Impressive as it is in flagging up the complex stylistic differentiations that operate within many clubbing cultures, Thornton's project leaves little room for, and thus provides few glimpses of, for example, the complex interactional demands of the dance floor in also contributing towards the constitution of 'cool' and the development of notions of belonging. In particular, Thornton says very little about the imaginative–emotional constitution of clubbing for – and usually by – the clubbers.

The close connection between the use of recreational drugs, such as ecstasy (MDMA) and speed (amphetamine), and the practices and spaces of clubbing is stressed in the series of texts edited by Nicholas Saunders (1994; 1995; 1997). Within Saunders' texts are contained valuable contributions about the links between drug use and clubbing by Mary Anna Wright. Again, however, and in a somewhat similar way and for similar reasons to Thornton, while Wright (1995) touches upon the nature of the experience, she is primarily concerned with the history of and current musical genres constituting the clubbing experience. Because Wright is concerned with the contention that dance cultures may represent 'social movements', she fails to engage in any detail with the sociological and technical aspects of the clubbing experience.

Thus, while these and other relatively broad discussions of clubbing, raving and the media treatment of these leisure pursuits are useful in an evolving understanding of the many aspects of clubbing, they provide more of a background to, than an insightful conceptualisation of, *the practices and spacings constituting clubbing*. They neglect the intricate demands and techniques of clubbing, tending rather to focus upon what Michel Maffesoli (1996a) might call an 'ocular understanding' of clubbing – an observed yet distanciated knowledge in place of a tactile understanding premised upon 'being-togetherness' and 'being-in-touch'. A handful of recent books, papers and Ph.D. theses that deal with more historical, practical and technical facets of clubbing are now being completed and in some cases published[10]. These are starting to become a critical mass in the process of improving our understandings of clubbing. Yet, when one considers that up to 500,000 young people enjoy the clubbing experience in Britain every week – more than all those who enjoy spectator sports, theatre, live music, comedy and cinema combined (Mintel, 1996; Thornton, 1995) – it is incredible that so little work has been completed on what the clubbers actually do when they go clubbing, and why. Furthermore, certain texts on the changing social and cultural meanings of contemporary 'youth' have started to use the languages of club cultures – occasionally quite literally – in labelling contributions that are actually only partly about those club cultures and are usually concerned with 'youth cultures' more generally. Indeed, a recently published, so-called '*Clubcultures Reader*' (Redhead, 1997) contained only seven contributions (out of sixteen) that actually discuss clubbing or raving, and some of these seven only broach the subject tangentially. To date there has not been a comprehensive attempt to provide an understanding of clubbing as an *experience*, involving practical techniques that are blended with emotional and imaginary understandings.

Shut up and dance: re-thinking resistance[11]

An apposite example of how the paucity of work upon the actual processes and practices of being young can affect our understandings more generally may be found within recent critiques of the CCCS studies of 'youth resistance'. The notion of 'resistance' as it was used in the CCCS debates about young people and their cultures was often simplistic and totalising, taking little account of young people's *imaginative and practical* construction of their own experiences, and instead privileging only the macro-political dimensions of so-called 'resistant' behaviour. The Centre for Contemporary Cultural Studies envisioned a variety of youth cultures as actively engaging in *rituals of resistance* (Hall and Jefferson,

1993). These rituals included elements of style, music and dance, as well as verbal procedures which implied an antagonistic relationship with those in power (Jackson, P. 1989). This resistant attitude to 'power' was seen largely as a response to the apparent *powerlessness* of the young people, who were starting to take on many of the responsibilities of adulthood, but who apparently had yet to experience any of the privileges. Resistance, then, was understood by the CCCS as the obvious reaction of young people to their social positioning. But were these rituals what the young people actually did, or were they representations of what the young people did as seen through the eyes of social scientists? Questions such as these were rarely asked, let alone answered.

Angela McRobbie (1994b; 1994c), one of the original members of the CCCS from the 1970s, makes this point about simplistic notions of resistance in her writings on the cultures of being youthful. Resistance, McRobbie proposes, might be 'down-sized' from the mega-political status that it enjoyed during the CCCS years particularly, and instead conceptualised as existing at the 'more mundane, micrological level of everyday practices and choices' (McRobbie, 1994b: 162). By doing this it is possible to see how a (sub)cultural identification, or sense of transitory tribal affiliation (Maffesoli, 1988b; 1995), might provide 'a way of life' in the most literal and humble meaning of the term (McRobbie, 1994c).

Of course, as McRobbie is quick to add, all youth cultures, inasmuch as they stake out a space, an 'investment in society', are inherently political (McRobbie, 1994b: 156). Yet, this does not necessarily mean that all youth cultures are political in a *mega*-political, 'anti-authority' sense, although some do appear to be so[12]. The practices of youth cultures can be as much about *expression* as about resistance; as much about *belonging* as excluding; as much about temporarily *forgetting* who you are as about consolidating an identity; as much about gaining *strength* to go on as about showing defiance in the face of subordination; and as much about *blurring* boundaries between people and cultures as affirming or reinforcing those boundaries. The extent of these alternative possibilities has not always been reflected in the literatures and debates. Once again, this lack of texture in understandings of young people partly results from the practices, spacings and timings of the ways in which these young people live their lives – the *process of being young* – being overlooked. This neglect of *process* is a more general feature of our understandings of the ways in which we experience many facets of our social lives. Indeed, this neglect of process is a point that features centrally in my second starting point.

A first starting point, then, are the specific weaknesses in existing accounts of young people's leisure and pleasure, and particularly a neglect of the *nature and spacings of young people's practices*. Specifically, I am interested in progressing what I argue are simplistic notions of 'youth resistance' by concentrating upon the nature of the vitality that it appears may be experienced in just one context of contemporary leisure: clubbing.

Starting point two

Geographies of experiential consuming

A second starting point for 'The night out' are approaches to and understandings of forms and practices of contemporary consumption, or 'consuming'[13]. As a 'central element in the way we all experience the world' (Sack, 1992: 1) consuming undoubtedly represents 'one of the focal concerns within the social sciences in the 1990s' (Gregson and Crewe, 1997a: 242). Countless histories and commentaries on both the rise and extent of all manner of consumption activities and consuming practices, from car-boot sales to white-water rafting[14], appear at times to have been themselves submerged beneath the flood of treatises, papers, books, journal articles and arguments through which these commentaries have been analysed, dissected, reformed and critiqued[15]. In short, it would seem that the 'subject of consumption' (Shields, 1992a) has been done to death, revived at the last, and then done to death again.

This text is not the space in which to re-open or review these debates and questions in detail. However, as I infer through the contexts of the 'The night ahead', there are significant practices and contexts of consuming within our everyday lives that have been neglected or overlooked completely in the academic scramble for pastures new. These practices and contexts are critically implicated in processes of identity and identification formation and amendment, of socialisation and sociality and of the negotiation of difference and otherness, and thus any conceptualisation of these practices and contexts may assist in unpacking an understanding of clubbing through 'The night out'. Their neglect is due both to what I argue is the rather narrow and limiting definition of what 'counts' as 'consumption' or 'consuming', but also to what represents an inability or reticence on behalf of the researcher of contemporary consuming to access these significant areas and practices. More specifically, then, I want to stress the centrality of what I am calling 'experiential consuming' – or the consuming of experiences – in contemporary social life. In particular, I want to do this in the context of group interactions, crowd activities and identifications. Before I address this in some detail, however, I first skip quickly through a number of more general weaknesses in the existing debates and literatures about the nature of contemporary consumption.

As part of the clamour to write about aspects of the worlds of consuming, to which I have just alluded, the geographies – or spatialities – of consuming are receiving increasing attention. Again, the directions and deficiencies of this work have received plenty of exposure and need no introduction. However, a number of key points that have a direct bearing upon the notion of experiential consuming that I outline later bear repeating. First, while studies are starting to recognise the crucial spatial element of consuming, most studies have lacked any engagement with the historical dimensions of this consuming (Jackson, 1995; Glennie and Thrift, 1992; Sack, 1992). The overwhelming focus has rather been on casting our gaze forward to the shape of emerging societies of so-called 'postmodern consumption' (Bauman, 1995a; Featherstone, 1991; 1995). Furthermore, numerous studies have promoted the supposed authority of the contemporary consumer (Abercrombie, Keat and Whitley, 1994; Featherstone, 1991) and the sovereignty – even the freedom – that the consumer now purportedly enjoys in the market-place. Yet, scant attention has been paid to the actual processes and practices of this supposed freedom – a freedom, it seems, that is in fact much less about free choice than the inhabitation and negotiation of existing geographical worlds and the impacts of lifestyles and processes of stylisation upon individuals (Bauman, 1988; Bourdieu, 1984; Kellner, 1983; Sack, 1992; Warde, 1994a; 1994b). This much-vaunted supposed freedom of the consumer has segued with another burgeoning area of study – or at least writing – which has been concerned with the tactics of resistance that one may employ at times, supposedly unwittingly, as a consumer. These tactics of resistance have often been cast as ploys that one may use to subvert the imposition of power by and through the capitalist system, particularly in the contexts of working and often using the example of the spaces of contemporary cities (de Certeau, 1980; 1984; Kellner, 1983; Scott, 1985; 1990).

The frenzied search for new forms and new practices to study has resulted in important aspects of the geographies of contemporary consuming being over-looked in favour of a handful of by now familiar, but nonetheless still interesting and evolving sites, practices and moments. These favoured sites include the mall[16], the department store[17], the fair/exhibition/carnival[18], and latterly, the supermarket–kitchen–restaurant nexus[19]. Within this ever-growing body of work, the overwhelming concentration upon the post-war cultures of mass consumption in America and Britain, and the rise and rise of advertising[20], processes of commodification and stylisation as part of those cultures, has eclipsed the perhaps less obvious and certainly more elusive geographies. These are geographies that are not about commodification in quite the same terms, but which are nevertheless integral to – and, particularly at the level of the consumer, largely constitutive of – wider yet equally significant processes of identity and identification formation.

In short, and following Jackson and Thrift's (1995) review, I would suggest that the deficiencies of many contemporary approaches to understanding consuming cultures might be summed up in terms of a general but far-reaching reluctance that geographers, as well as others, have had to engage with the practices and social spaces of sociality (see p. 24) and the constitution of consuming experiences by individuals and groups more broadly. This is a further outcome of the ethnographic inertia of geographers, and tallies directly with the general paucity of detailed understandings of young people, their social lives, practices and spaces that I flagged up earlier. We know staggeringly little about the interactional order (Goffman, 1959) through which social life is played out on a micro, everyday, at times mundane and often non-rational level – what Danny Miller (1995) refers to as the ways that 'ordinary people' actually live their lives. This interactional order, and the practices and spaces of sociality through which it is formed, reformed and navigated, is at the core of the consuming geographies that I am highlighting during 'The night out'.

Consuming geographies – experiential consuming

Countless forms of consuming do not involve the purchase of objects or tangible services, but are instead premised upon an experience during and after which nothing material is 'taken home' – only an experience held subsequently as a set of memories. Even shopping, the form of consuming seemingly dearest to many theorists of consumption (presumably because it is what they do most often themselves) is at least partially constituted through social processes in which the purchase of objects is actually secondary to the thrill of the throng. Shopping in an empty shopping centre, for example, would for most people be as strange as not shopping at all. Yet, few attempts have been made to extend this recognition into less obvious or perhaps methodologically more problematic areas or modes of consuming. In particular, few attempts have been made to explicitly look at the active creation of the consuming experience by the consumers themselves[21].

This general absence is perplexing. An increasing emphasis on consumer reflexivity and the associated growth in both identity politics and individuation processes that have accompanied what Glennie and Thrift (1996: 234) call this 'reflexive turn' have resulted in a heightened role for the practices constituting sociality, the role of identifications and the sites and spaces of social centrality that promote them. Thus, while the processes and practices of many forms of socialised and group consuming – holidays, the leisure centre, shopping, cruise ships, sports spectatorship, *Centerparcs*, wine tasting, bingo, gambling, 'dangerous' sports, paint-balling, snow-boarding, working out at the gym and, er...clubbing (to name just a handful) – are seemingly more popular then ever, and at the least are of heightened prominence in society, this popularity is not reflected in research agenda.

From their relative absence one might assume that these forms and spaces of consuming and their constitutive practices are either less easily accessible to the researcher, and/or unworthy of or unsuitable for study. Colin Campbell (1995) appears to concur with the latter of these assumptions in voicing his concern over the extent and boundaries of current work on consumption. Campbell proposes that the consumption of 'spectator sports' and the 'arts' are steps too far: 'the suspicion must arise that in attempting to encompass so large and diverse a range of topic areas, the sociology of consumption may lose whatever meaningful or distinctive character it is in the process of developing' (1995: 111). Yet, only a few pages further on, Campbell presents his case study of tourism, which he suggests is 'modern consumption illustrated' (ibid.: 117), in extending his theory that modern consumption is about the consumption of 'novelty', with the 'essential activity of consumption' being the *'imaginative pleasure-seeking* to which the product image lends itself' (ibid.: 118, my emphasis). This would seem a confusing contradiction. For example, conceptually how different are 'tourist' experiences of France on the one hand and World Cup football fandom in France on the other? Or going to an art gallery to view pictures and then going to the gallery shop to purchase a print?

Campbell contends that modern consumption might best be understood as a form of imaginative hedonism (see Campbell, 1987) in which anticipation is the main source of pleasure and actual consumption is likely to be 'a literally disillusioning experience' (1995: 118). Yet Campbell's thesis fails to recognise the value of, practices constituting and pleasure gained through experiences as varied as drug use (including the use of alcohol), dancing, eating and hill-walking. While the power of the imagination in contributing towards the allure of certain forms of consuming cannot be denied – and in fact, as I argue through 'The night out', must be central within our conceptualisations – the actual practices of consuming themselves are surely *also* implicated as fundamental? On a very basic level, for example, the *imaginative* construction of consuming is inextricably tied into the *practical* constitution of consuming through the use of the body.

Ignoring the imaginative and practical constitution of many consuming experiences has meant that these experiences have often remained disembodied, decontextualised and bereft of emotions, imaginations and sensations. One result of the empirically restricted boundaries around studies of contemporary consuming has thus been a rather limited conceptual understanding of how and why people consume what they do, where, with whom and how this intersects with other facets of their lives. I am, of course, not attempting to comprehend *all* the various forms of collective, experientially based consuming. Rather, I am highlighting one form – the practices and spatialities of clubbing – in a quest to progress and give texture to a number of current understandings of consuming more generally.

Thus, a second starting point for 'The night out' are existing geographies of consumption, many of which appear narrow and somewhat de-humanised in their understandings of how consuming is comprised through both practical *and* emotional facets. Excessive concentration on certain sites and forms results in other practices and spacings of consuming being ignored. In 'The night out' I am interested in exploring notions of what I refer to as 'experiential consuming' in which the concurrent *emotional and practical constitution of the clubbing experience* by the clubbers themselves is based upon and within notions of group togetherness, of the crowd, and of the technical demands, and opportunities, of clubbing.

At the centre of consuming experiences such as clubbing, in which nothing material is purchased, are the practices of sociality – how the cultures of clubbing are actually constituted. These practices of sociality may be conceptualised through what can be referred to as a 'performative lens'. These two points together comprise my third starting point.

Starting point three

Geographies of sociality and performativity

As a third starting point I am interested in the relationships between processes of consuming, crowds and co-presencings, and the performance of identities and identifications in those crowds through the practices, spacings and timings of sociality. I now introduce the notion of sociality in some detail and link it explicitly with notions of experiential consuming. I then introduce the notion of performativity and suggest how crowd-based consuming experiences might be understood as simultaneously *expressive* and *constructive* of self. I finish by proposing that through conceptualising consuming practices and spacings through this performative metaphor it may be possible to gain a number of valuable insights into the relationships between practices, identities and contexts of identifications.

Sociality

Sociality may be defined as the common sense or human nature that underlies the more formal aspects of social life (Maffesoli, 1995). Sociality can thus be understood as the 'softer' side of sociability. Sometimes seemingly invisible, at times secretive, and often elusive, sociality has been described as the dark underbelly of society and society's norms, mores and civilising processes (Maffesoli, 1993a). Sociality constitutes the 'world-that goes-without-saying' (Amirou, 1989: 119), where ' "what goes without saying" makes the *community*' (Maffesoli, 1993a: xiv; emphasis in original). For these reasons, the practices of sociality have been described as the *glutinum mundi* of social existence (Maffesoli, 1995; Shields, 1992a). In their discussion of the role of the 'throng' or crowd in the shopping experience, Glennie and Thrift (1996) define sociality as:

[T]he basic everyday ways in which people relate to one another and
maintain an atmosphere of normality, even in the midst of antagonisms
based on gender, race, class or other social fractures: what Shields calls
'the connecting tissues of everyday interaction and co-operation'
(Shields, 1992a: 106).

(Glennie and Thrift, 1996: 225)

The practices that comprise sociality consist of ways of dressing, spoken and
unspoken languages, traditions and customs, myths and folklore, and the sharing
of styles, knowledge and passions. The practising of sociality requires skills,
knowledge and competencies, all of which need to be acquired, regulated and
honed, usually through repeated participation within the context of the sociality
concerned, and blended into a form of self-administered 'education' (Gregson
and Crewe, 1997a; 1997b). A sharing of these practices of knowledge acquisition
can act to bind groups of disparate individuals, as well as to provide the basis for
processes of distinction and the construction of boundaries between social groupings. In contrast to previous conceptualisations of sociality (such as those
elaborated by Goffman and Simmel), more recent work tends to emphasize the
weak and fluid nature of this form of social relation. Glennie and Thrift (1996:
226), for example, suggest that the notion of sociality 'is worth examining at the
present time because of important differences between early and current formulations of the concept'[22].

The growing number of texts by Michel Maffesoli (1988b; 1993a; 1995; 1997)
are especially provocative in this respect and provide a wealth of points of departure for any study of contemporary formations of sociality[23]. At times
frustratingly off-hand, vague and nebulous, the *oeuvre* of Michel Maffesoli provides
a rich and provocative commentary upon the changing dynamics of social life and
especially the 'rise of the new consumer society' (Maffesoli, 1997). His project is
the development of a sociology oriented towards a recognition of *sociality* as a – if
not *the* – central aspect of social life (Maffesoli, 1995) and a sociology that highlights the changing structures of everyday social interactions within this social life
(Shields, 1991b). As part of this developing sociology, Maffesoli is interested in
highlighting what he describes as (alluding to Weber's (1976) notion of 'dis-
enchantment') the 'reenchantment of the world...by means of the image, myth,
and the allegory...that characterise contemporary style' (Maffesoli, 1996b: xiv).
Through emphasizing the notion of identifications over that of identities, and
highlighting situations in which one's social environment can temporarily
submerge one's sense of identity, Maffesoli is concerned with social configura-
tions that go beyond individualism: 'the undefined masses, the faceless crowd, and
the tribalism within them' (Maffesoli, 1995: 9). Finally, Maffesoli privileges a role
for the crowd or what he terms a 'being-togetherness' (1995: 81). This

being-togetherness is based upon an 'empathetic sociality' (1995: 11), and the constant coming and going between these crowds or 'neo-tribes' (1988a: 145). Maffesoli uses this sociality as the basis for his conception of 'puissance' – or 'the will to live' (1995: 31) – that can energise individuals and in doing so act to bind more strongly these neo-tribal formations.

Maffesoli is keen to emphasize the everyday nature of the practices of sociality, as well as their fundamental presence within social life on a basic level that is experienced by all, and their flexible, negotiable and overlapping form. For Maffesoli, the practices of sociality should be understood more as the fabric of tactile or proximate forms of communality than of the strong 'contractual' form of community that characterises 'the social' (Maffesoli, 1995). While the former are premised upon fluid, unsettled and unsettling social crystallisations and dispersals, the latter are based upon clearly defined lines of interaction (Maffesoli, 1993a). Thus, for Maffesoli, sociality is less about rules and more about sentiments, feelings, emotions and imaginations; less about what has been or what will become than what is – the stress is on the 'right now' and the 'right here'.

For all the delicate, occasionally polemical and even obtuse arguments in Maffesoli's work, a major weakness remains that it almost completely lacks empirical contextualisation. Of course, this is also one of the seductive features of his writings, which are so clearly and unashamedly theoretical. For example, although provocative and useful in evoking some contemporary forms of temporary community and the sociality through which such belongings are established, Maffesoli's 'neo-tribes' thesis fails to evoke the demanding *practical* and *stylistic* requirements and competencies that many of these communities demand, and through which many of them are constituted. Affiliation and a sense of identification are inferred as being almost a matter of course, as openly 'up-for-grabs'[24].

As I unpack in more detail in 'The night out', the practices of sociality that constitute clubbing provide an empirical texturing to some of Maffesoli's notions and conceptions. To give some examples: the relatively weak sense of a fixed *order* in clubbing in no way precludes a strong sense of dynamic *orderings* (Hetherington, 1997; Law, 1994); the mooted ease with which individuals supposedly move between, and even within, Maffesoli's 'neo-tribes' is challenged; 'access' to clubbing crowds is clearly not 'open to all'; and the idealism and utopianism, which at times taints Maffesoli's notions of 'being-together' and 'puissance', are problematised and given practical context. That said, what makes the constituent practices of sociality relevant to the evolving notion of experiential consuming that I sketched out earlier is that they highlight the dynamism – or *process* – of consuming. Furthermore, and crucially, it appears that these practices of sociality are often themselves what is being consumed. The notion of performativity is one approach to understanding the relationships between these practices of sociality and the identities and identifications of the individual consumers.

Performativity

The practical negotiation of crowd membership or affiliation involves processes of communication that are always dialogical, that is: they are two-way processes (Canetti, 1973; Le Bon, 1930; Maffesoli, 1988b; 1996b). These processes of communication that take place within a crowd, and can act to bind that crowd, are what Boden and Moltoch (1994) call 'body talk', Mauss (1979) labels 'body techniques', and Goffman (1963) refers to as 'body idiom'. The 'dance of life' (Boden and Moltoch, 1994: 260) through which self is expressed and constructed, and similarly others within the crowd are understood, is relevant to the notion of sociality outlined earlier for at least three important reasons. First, individuals within a society or culture often attach similar meanings to physical appearance and bodily actions such as gestures and facial expressions. These are used as a common means of understanding embodied information. Second, the body generates these meanings even when an individual is not consciously articulating them. Constant attention is required for the successful management of the meanings projected by the body and even this is not always sufficient. For example, few people can control their sense of embarrassment, no matter how hard they try – it just happens (Goffman, 1959; Shilling, 1993). Third, sociality comprises a learned set of practices and techniques. The consuming crowd can function ' as a kind of classroom, as a means of reflexivity' (Glennie and Thrift, 1996: 227) through which individuals learn not just about commodities and their uses and meanings, but also about styles of self-presentation, bodily techniques of expression, and associated notions of competency or even expertise. The consuming crowd becomes a potentially important context for the construction, transformation and expression of the 'imaginary orders that constitute people's self-projects' (Glennie and Thrift, 1996: 235). In these ways sociality can be further understood as comprising performed, and thus learned, practices through which identity can be constructed and notions of belongings may be established. It is useful to develop briefly this notion of performativity through a brief review of some of the associated work of Judith Butler (1990a; 1990b; 1993) and Erving Goffman (1959; 1963; 1967; 1968; 1969; 1971).

In her notion of gender as a performative act, Judith Butler (1990a) is inspired by Simone de Beauvoir's (1972) maxim that 'one is not born [as], but rather becomes a woman'. Butler criticises the commonly held idea that femininity and masculinity are merely the cultural expressions of the material maleness or femaleness of the body, instead proposing that the notion of gender might be better understood as performative (Aalten, 1997; Butler, 1990a). For Butler,

gender is constituted through the on-going and repeated stylisation of the body
rather than through any notion of biological facticity or cultural understanding.
Thus, she proposes that bodily gestures, movements and enactments effect the
illusion of what she refers to as 'an abiding gendered self' (Butler, 1990b: 270).

An important strand of Butler's work is her emphasis upon the connectedness
of the body and the subjectivity of the individual[25]. Butler develops this point in
detail:

> [A]cts, gestures, and desire produce the effect of an internal core or
> substance, but produce this *on the surface* of the body, through the play of
> signifying absences that suggest, but never reveal, the organizing prin-
> ciple of identity as a cause. Such acts, gestures, enactments [...] are
> *performative* in the sense that the essence or identity that they otherwise
> purport to express are fabrications manufactured and sustained through
> corporeal signs and other discursive means.
>
> (Butler, 1990a: 136; emphasis in original)

For Butler, then, the gendered body has no 'ontological status' (ibid.: 136)
outside the performative actions which that body practices, where that performance is rooted within and routed through the bodily surface of the interior and
organising core: 'gender becomes a performative and regulatory practice'
(Pinnegar, 1995: 51). This regulatory practice is one 'that requires constant repetition' (Cream, 1995: 38). Through developing the example of 'drag', Butler goes
on to claim that it is clearly possible for appearances to be an illusion, that the
supposed distinction between inner and outer may be 'mocked', 'subverted' or
'parodied', that the notion of a 'true gender' identity can be displaced, and thus
that notions of identification are always performative (Butler, 1990a: 136–41)[26].

Parts of Erving Goffman's much larger and earlier *oeuvre* can be used in
contextualising Butler's at times abstract yet conceptually fascinating notions.
Goffman is concerned first and foremost with the study of the face-to-face interactions that occur during situations of co-presence: 'the ultimate behavioral [sic]
materials are glances, gestures, positionings and verbal statements that people
continuously feed into the situation, whether intended or not' (Goffman, 1967:
1; see also Goffman, 1969). Goffman develops this interest through further
concerns with the ground rules and associated orderings of behaviour through
which social spaces are constituted (Goffman, 1968; 1971), with the 'entry qualifications' that distinguish public from private spaces, and with the 'multiple social
realities' through which the same physical space may be experienced (Goffman,
1963).

Perhaps the best-known element of Goffman's work is his interest in developing an understanding of social life through the 'dramaturgical metaphor'; that

is, Goffman uses the notion of theatrical performance as a perspective upon social life more broadly (Goffman, 1959). It is through the development of this dramaturgical metaphor that Goffman is also most clearly concerned with notions of *spatialisation*. This interest in spatialisation manifests itself most obviously in Goffman's notions of 'backstage', 'frontstage', 'settings' and 'the outside' (Goffman, 1959), but also in more complicated notions of 'open' and 'closed' regions and 'containment' (Goffman, 1963), and 'use space', 'personal space' and 'possessional territories' (Goffman, 1971).

One of the major criticisms of Goffman's approach is that he tends to neglect aspects of *social identity*, such as ethnicity, gender and sexuality, except in much broader terms, for example when he develops his notion of *Stigma* (1968) (Kendon, 1988). Goffman instead concentrates on developing understandings of the timings and spacings of bodies within social encounters (Lemert and Branaman, 1997). Two criticisms of Judith Butler's notion of performativity are that she discusses nothing *except* gender as an aspect of social identity and that her analysis is abstract and more specifically a-spatial (Mahtani, 1998). Thus, through fusing Goffman's recognition of the role for territorialisations and regionalisations with Butler's notion of social identity and self being performed concurrently, we arrive at a more textured and practically useable conceptual lens through which to approach notions such as belongings and the practices of sociality that sustain them. Most significantly, through alloying together these two approaches to performativity we can improve our understandings of how the consuming experience of the crowd can be simultaneously expressive (Goffmanian) and constructive (Butlerian) of self. Thus, by conceptually 'folding together' the practices and spacings that constitute processes of identity formation and identification formation, it is possible to open out new understandings of the relationships between being 'in the crowd' and being conscious of one's self. This fluctuation between experiencing social situations through notions of self and through being part of, and submerged within, the crowd is a further theme that runs throughout 'The night out'.

Performing consuming

The work of Goffman and Butler demonstrates that approaching social life as a performance in this way is, of course, not new. Yet conceiving of consumption as a *process* of consuming in which the consumers are actively *performing* their involvement across time and spaces, and through which aspects of both their identities and identifications are concurrently constructed, transformed and expressed, does provide a number of further advantages.

First, the consumers are not merely positioned as passive spectators, rather, they are understood as embodied actors, engaging with the consuming world as both spectators and audience simultaneously, and in many ways indistinguishably (Crang and Malbon, 1996; Sack, 1992). Second, this conception acknowledges the significance of consuming in terms of self-investment and the efforts, techniques and competencies constantly required for the maintenance and amendment of self-identity, as well as in the establishment and reinforcement of group identifications. Third, the crucial contextual stagings – what I refer to in 'The night out' as 'spacings' – of consuming and the relationships between these stagings are brought to the fore. Thus, notions of backstage and front stage, of display and concealment, of collusion and group work, of props and scenery (environment), and of the design and 'inhabitation' of sets and spaces, are prominent. Furthermore, notions of audience and performer, of the 'in' cast and 'outcast' – notions of coolness and distinction – of inside and outside (or spatial belongings), and finally of timing and routine, are highlighted. Fourth, approaching the processes and experiences of consuming through notions of performativity also foregrounds a sense of the structural power of *scriptings*, which intersects with notions of the '*habitus*' (Bourdieu, 1984)[27]. Consuming is, of course, not completely fluid and 'up-for-grabs', but is cut through by the surveillance practices of *directing* – or production and staging – as well as the tactical nature of the multiple interpretations, or ad libbings, of these scripts. Fifth, approaching consuming and its constitutive practices of sociality through the lens of performativity allows us to challenge the neat and oft-cited dichotomy between power and resistance – de Certeau's (1984) strategies and tactics, and Maffesoli's (1995) power and puissance spring to mind – through a recognition of the significance of notions of strength and vitality in everyday life, and their role as potentially desirable outcomes from a 're-reading' of the script (Crang and Malbon, 1996). These notions of strength and vitality, which Maffesoli approaches in his conception of puissance yet never quite hits directly, are curiously absent from the literatures on resistance. Instead, these literatures, as I indicated briefly in the discussion of the geographies of youth, seem intent on meta-politicising virtually every human action.

This third starting point, then, gives added texture to the first two starting points, providing the foundation of an approach to understanding clubbing which foregrounds the practical and emotional nature of the experience. At the same time, this last starting point provides an approach with which one can explicitly 'flag up' the relationships between the motions and emotions of clubbing, and the identities and identifications of the clubbers. The concern here is with the centrally significant yet presently largely unrecognised role for the *practices and spacings of sociality* in the constitution of these consuming experiences, and in the nature and (re)orientation of notions of resistance and vitality. Through approaching the relationships between this sociality and the identities and identifications of the consumers concerned at least partially through a performative lens

— a lens that prioritises the *concurrent development of identity and of identifications* within the crowd — I suggest that we can give vital contextual detail to our understandings of these social formations.

The starting points...and clubbing?

'The night out' is structured through four thematically ordered sections, each of which provides a progressively more deeply textured understanding of clubbing as an experience. Broadly stated: the four sections that now follow relate the three starting points that have just been outlined to notions of style and belongings, to musical and dancing crowds and the technically demanding nature of crowd belongings, to moments of ecstasy and euphoria and the associated use of recreational drugs, and to alternative approaches to understandings of resistance — approaches that prioritise the role of 'play' in the experiencing of personal and social vitality.

RESEARCHING CLUBBING

Clubbing is usually subdivided by clubbers and the clubbing media according to sexuality, age and location into types or strands, such as mainstream, gay, student, S&M, indie, 'local', as well as being differentiated into musical genres, such as house, techno, drum'n'bass and big beat to name only a handful of genres that might themselves be further sub-classified. Although each of these genres will be constituted through differing timings, spacings, musics, styles and technical demands, in attempting to present a new understanding of the practical and emotional constitution of clubbing there is clearly no way to cover all these genres and sub-genres within a project of this size. Thus, in selecting a form or genre of clubbing, and the associated clubbers and clubber stories upon which to base the understanding of clubbing that follows, I restricted my research to a relatively narrow genre of clubbing[28].

First, I chose to access what I perceived as the largest *general* group of clubbers; that is, I did not explicitly set out to include indie music clubs, S&M clubs or suburban Ritzy-type clubs, to give just three examples of sub-genres. I also selected clubbing experiences that I believed I would have the least problems in accessing and in gaining real understandings relatively quickly. I wanted and needed to be able to empathise with the clubbers who had agreed to help me, and I wanted to 'be in touch' with them as far as I could be. Primarily, I was keen to attain, as near as possible, an 'equal relationship' between myself as interviewer and the clubbers as interviewees. I did not want to be seen by the clubbers as more of an 'outsider' than was partly inevitable. My own background as a clubber was, I believe, crucial in establishing my credentials as someone who was both genuinely interested in and could readily empathise with their experiences rather than merely someone who just happened to be 'doing a project' on nightclubs as his 'job'. Second, I specifically selected clubbers who lived in or very near London and who went clubbing regularly in central London. This was for reasons of ease of access, both for the clubbers and for me.

I wrote letters requesting clubber volunteers to all the major clubbing, 'style' and music magazines in Britain, of which only two printed them: *The Face* and *i-D*. I made a similar request for help on two Internet discussion groups[29]. In total I received over forty offers of help from places as diverse as Russia, Brazil and, er...Dartford. Of these, I went clubbing with and interviewed eighteen clubbers over a period of a year[30]. Of those who offered their help, I went clubbing with nearly all with whom it was practically appropriate to go.

Clubbers were usually interviewed twice, with the second interview happening after we had been clubbing together. Both interviews were very much 'conversational' in style and I avoided interview schedules, although all interviews were taped. The first interview was designed to achieve three main goals: to put the clubber at ease while also explaining clearly and fully in what ways I was hoping for help; to begin to sketch in details of the clubbers' clubbing preferences, motivations and histories; and to allow me an opportunity to decide how to approach the night(s) out that I would be spending with the clubber – what shall I wear? what kind of club will it be? how central might drugs be? This preparation was crucial to the success of the nights out.

The second interview provided a forum for what was invariably a more relaxed meeting than the first interview, given that we had spent at least one evening together in the interim period. Indeed, a sense of increasing informality is perhaps the major benefit of this type of serial interviewing. The main content of the second interview consisted of comments, discussion and questions about the club visits we had made together, and the nature of the night out as an experience. In the latter half of these second interviews, discussion occasionally diversified in scope to cover wider aspects of the clubbers' lives: their relationships to work or study, their relationships with friends and loved ones, their hopes and fears for the future and their impressions of a social life beyond and after clubbing.

These introductions are now almost complete, and the 'The night out' beckons. However, as a final thought and on a more personal note, I should add that throughout the year of the main research the clubbers with whom I went clubbing and who agreed to be interviewed were without exception unbelievably helpful. During these twelve months, I spent almost 150 nights out – many of these were the best nights out I have had.

> But now it is evening. It is that strange, equivocal hour when the curtains of heaven are drawn and cities light up... Honest men and rogues, sane men and mad, are all saying to themselves, 'The end of another day!'. The thoughts of all, whether good men or knaves, turn to pleasure, and each one hastens to the place of his choice to drink the cup of oblivion.
>
> (Baudelaire, 1964)

Part Two

THE NIGHT OUT

Night-life...night-club...night-bird. There is something about the word 'night', as about the word Paris, that sends through some Englishmen a shiver of misgiving, and through another type a current of undue delight. The latter never get over the excitement of Sitting Up Late. The others see any happening after midnight – even a game of snakes and ladders – as something verging on the unholy; as though Satan were never abroad in sunlight. A club they can tolerate. Call it a night-club, and they see it as the ante-room to hell.

(Burke, 1941: v)

GETTING INTO IT, FEELING
PART OF IT

club – (often followed by *together*) to gather or become gathered into a group; to unite or combine (resources, efforts, etc.) for a common purpose [C13: from Old Norse *klubba*, related to CLUMP).

(*Collins Concise Dictionary*, 1995, third edn)

INTRODUCTION

This first section of 'The night out' is about belongings and distinctions. Notions of belonging and distinction are central to the clubbing experience, and, indeed, are crucially implicated in both pre- and post-clubbing times and spaces. In particular, clubbing is heavily entangled with notions of identification and style, even if only in the form of a conscious rejection of an organising style or 'subculture'. The word 'clubbing' is itself connotative of membership and belongings. Developing this idea, throughout this initial section I present and critically discuss a number of facets of *pre-clubbing*, which in many ways precede and provide important foundations for subsequent stages of 'The night out'. I am interested in much more than merely flagging up how the clubbers to whom I talked first began clubbing. Although significant, I want to go beyond these relatively simple tales of 'getting started' to look in some detail at issues of belongings as evoked and experienced by the clubbers who assisted me in the research for 'The night out'.

More specifically, this section is about drawing out some of the relationships between notions of stylisation or 'coolness' and the practices and negotiations of group belongings and identifications. I develop this interest in belongings and distinctions in exploring three main themes that orbit around these early stages of the night.

First, I explore some of the reasons clubbers give for *going clubbing*. After discussing this idea of alternate modes of social interaction and illustrating the point through the stories of clubbers, I then develop this exploration through introducing the notions of identifications and belongings. Identifications, I argue, do not simply replace identities. Rather, the identifications that may be experienced through clubbing – and especially the sensory overload of the clubbing crowd – temporarily supersede certain facets of individual clubbers' identities, although in most instances only for fleeting moments.

Second, developing the initial points about belongings, as well as spinning off numerous clubber stories about issues of style or 'cool', I outline a brief conceptualisation of the clubbers' relationships to issues of being cool. Stylisation and constructions of coolness appear central to processes of identification, with notions of the 'in' crowd and 'out' crowd, and choice of club night being crucial and, it appears, always relative. What is 'cool' for one clubber may be the epitome of 'un-cool' to another.

Third, and again following on from these first two themes, I turn explicitly to clubbers negotiating entry to the club as an important influence on the constitution of these clubbing belongings and the crowds to which I turn in the second section. At times, and for some more than others, entry appears tricky, with aspects of identity seemingly easier for some to relinquish than for others.

GETTING INTO CLUBBING

Questions of stylisation and the management and presentation of coolness, club choice, club access – the vital negotiation of 'the door' – and social exclusivity (being refused entry) will be broached and deepened in the latter half of this section. For now, I draw out three related aspects of clubbers' reasoning for going clubbing. First, I highlight the notion of clubbing as constituting a quite different form of social space within the city to, say, the 'streets outside'. Second, in highlighting what are apparently the different codes of social interaction that clubbing demands and allows, I look at the example of women's experience of clubbing in comparison to their experience of other, perhaps more oppressive and intimidatory, social spaces. Third, and building on these first two points, I highlight the clubbers' articulations of some form of need or desire for a sense of belonging, and further unpack the notion of identifications which I briefly mentioned in 'The beginnings' as being potentially helpful in understanding some of these impulses towards clubbing.

Being in the city yet being out of the city

MARIA: You know, when you live in a city like London where everything goes pretty much...I was desperate to get away from Southend – I really wanted

to come to London, ummm, as it turned out, which is a bit weird. I come
from a village just outside Southend and my parents moved there when it was
quite small there, a lot of people in the village know my parents because they
were involved with a lot of things when I was younger, umm…I was sort of,
in a way, a little bit of a celebrity when I was younger because I used to do a
lot of swimming and people knew me from my swimming and so everybody
knew me and it was nice to come somewhere where no one knew the fuck
who you were and no one gave a shit anyway, so [yeah, right] it was great,
yeah. I don't like the – I now don't like the anonymity of it, well sometimes
it's nice but erm, it would be nice to – I mean people can be very unfriendly
in London, but that depends upon how you take it because I do like that,
because sometimes I don't want to talk to people so I don't make the effort
and other days I'm feeling good and happy and I'll…I'll just start talking to
somebody on the tube or I'll walk into a shop and just have a conversation
with somebody, so it's how…it's nice because I can have it both ways. I was
interested in sort of going out clubbing here [in London] because I knew that
I didn't like what was happening in Southend that I had experienced…I
suppose I felt quite naive, and erm, quite sort of green as well, I definitely
felt very uncool and untrendy, umm, you know?

Maria's differing relationships with the place from which she originally comes (a
village near Southend) and with London begin to emerge in this extract. First,
London seems to have released her from a stifling sense of familiarity with others
who lived nearby. Second, the apparent anonymity of London – including its
clubbing experiences – was, for Maria, potentially both a pleasant and an
unpleasant experience ('I don't like the anonymity of it, well sometimes it's
nice') – she doesn't have the obligations of having to interact with those she is
with, yet she can if she so desires. Third, and related to this, during her early clubbing experiences in London Maria felt 'quite naive…definitely very uncool and
untrendy' compared with clubbing in Southend. The city is a qualitatively and
socially different experience for Maria.

This city life that Maria evokes through her story appears to be seductive yet
also, at times, overpowering. This paradoxical quality of the city is one with which
many writers have grappled and have attempted to distil through their writings. A
myriad of possibilities, lifestyles and activities are on offer in the city in which 'we
are barraged with images of the people we might become' (Raban, 1974: 64). The
'wrap-around reality of contemporary metropolitan life' (Chambers, 1990: 60),
structured through processes of time-keeping, punctuality, spatial ordering and
the 'organisation' on the one hand (Simmel, 1950), yet appearing more as 'a
maniac's scrap-book, filled with colorful entries which have no relation to each
other' on the other (Raban, 1974: 129), creates a sensory and stimulatory

environment of intensity and complexity. At times this can be pleasurable, but it can also result in 'sensory overload' and unpleasant notions of social exposure – of being alone amidst a city of strangers. City life, claims Bauman (1995a: 126) 'is carried on by strangers among strangers', and it is this continual and often stressful presence of strangers that can result in the impulse for experiencing an alternate atmosphere of belonging or identification. 'Modern living means living with strangers, and living with strangers is at all times a precarious, unnerving and testing life' (Bauman, 1993: 161).

In his musings on nineteenth century Parisian life, Charles Baudelaire was struck by the relief from their 'inner subjective demons' that citizens experienced in a city of passing encounters, fragmentary exchanges, strangers and large crowds (Sennett, 1990). For Baudelaire, this mode of experiencing the city of constant flux as pleasurable was perfectly captured in the figure of the *flâneur* – a citizen who takes visual possession of the city, wandering at will without a prescribed purpose other than exposure to diversity, gaining a form of melancholy pleasure through this experience (Wilson, 1995)[1]. However, for many if not most people in the contemporary city – perhaps un-schooled in the ways of the *flâneur*, perhaps merely '*flâneured* out', for even *flâneurs* need days and nights off – the social spaces of the city, in which relief from 'inner subjective demons' may be found, are usually less exposing, less structured through formalised codes of civility and much more constituted through the establishment and experiencing of intimacy than are the 'streets' of which Baudelaire writes (Bauman, 1995b; Sennett, 1990). These social spaces of alternate ordering are premised upon open expressions of spontaneity, emotion and conviviality rather their suppression or exclusion.

The presence of strangers, by definition, means that there also becomes possible the identification of those with whom one might establish a belonging, literally: a sense of identification or a bond of some form. The social spaces of the city are not only constituted through situations of anxiety and exposure, then, but also through spaces of identification, spaces that exist within and between the more anonymous spaces of exposure. These spaces of identification can provide opportunities for personal expression, conviviality and a going-beyond of the self as the individual is, usually both affectively and electively, subsumed within a larger social situation.

LUKE: In general, if the public is happy, open minded and cheerful, I get that feeling too. And if the crowd is nasty, shy and things like that, I have a boring evening. Just a little story about dancing together…I remember a night in another club, Shiva, a club with lots of dope and quite a lot of 'ordinary' people, and I was particularly out of my head on E, and suddenly there was a gabber[2] dancing in front of me, with a lighting stick, which he moved around my head. That was nice, we danced for a while and had fun together, though we wouldn't see each other while walking in the streets.

Of particular interest in this extract is Luke's suggestion that he and the 'gabber' would not interact in this way in a *non-club* situation (the 'streets') and in fact they 'wouldn't see each other'. This sensation of 'not seeing' others who are physically co-present is reminiscent of the 'scrutiny disguised as indifference' of which Lofland (1973: 178) speaks in her discussion of Erving Goffman's (1971) notion of 'mismeetings'. In cases such as these, claims Lofland, the 'trick' or the technique is to see while feigning complete indifference – to look at but also stare through simultaneously[3]. Luke thus contrasts the interactional ordering of clubbing with that of 'the streets'. A common feature of clubbers' accounts was the quite different set of codes of social interaction that clubbing appears to both demand and allow. Clubbing was described as being less about the *indifference* of the crowd – so beloved by Baudelaire, for example – and the *civility* of structured distanciation necessitated by the crowd, as discussed by Bauman, and more about the identifications possible with and through that clubbing crowd. This notion of belonging, and the associated sensation of liberation experienced through this interactional alterity, seemed to be particularly significant for some of the women clubbers that I interviewed.

Codes of social interaction – women, clubbing (and 'liberation'?)

CARMEL: If you're out, as far as the people I knew – blokes and girls included – you'd go out and the whole intention was to go out and get a bit out of your face and dance, sweat and have a laugh. Just having a laugh with none of those sort of social pressures I suppose [right]...blokes that come into the party might, or girls, that maybe were a bit insecure before that...it wouldn't matter in a party cos you could do exactly as you'd like, it didn't matter, and people didn't give a shit. For me guys do feel like that – they feel, y'know, that girl, she may look nice but she hasn't come out to get chatted up – she's come out here with her mates to get a bit out of it and dance so maybe I'm going to interrupt her night by trying to talk to her on that level because, I mean...to me it's just, it was just...if someone ever did come and talk to me it was like 'No!, I don't want to talk to you, this is a really good record – I want to go and dance!' and that's what I'd be thinking all night.

It is clear from this extract that Carmel goes clubbing primarily to dance, but I want to draw out three further and in some ways more subtle points. First, Carmel suggests that 'none of those sort of social pressures' are present, at least among her group of friends (the social pressures she mentions are presumably those that characterise other social spaces she experiences). Second, the absence of these conventional social rules and their replacement with another alternate ordering was recognised by the men as well as the women. Third, the quite detailed evocation of what a man might be thinking – 'she's come out here with her mates to get a bit out of it and dance so maybe I'm going to interrupt her night by trying to talk to her on that level' – suggests that Carmel has thought in some depth about how and why clubbing is particularly attractive to her, as well as how others understand it.

Of the clubbers that I interviewed, some of the women appeared to have been at first pleasantly surprised by and now currently value the different 'rules' and customs guiding or influencing forms of social interaction and inter-subjectivity in their clubbing experiences. During our discussions of their reasons for going clubbing, these women articulated these differences in interactional regimes between clubbing and other social spaces through talking about feelings of sexual liberation and a relative absence of often predatory men. The presence of the latter was viewed by many women, it appeared, as an unwelcome feature of other social spaces, such as bars and pubs[4]. Often, rather than clubbing being premised upon opportunities for sexual interaction, it appeared that it was dancing that was central. Thus, while women are undoubtedly still interested in 'looking good' – and will often want to appear 'sexy' and 'attractive' – while clubbing they might not feel the same pressures as they would feel in a pub situation, for example, to interact in a more concerted fashion with co-present men. This subtle difference is inferred in the extract from an interview with Carol and Lucy:

CAROL: You go out obviously, and you look around, d'you know what I mean? Especially when you're with your mates.

LUCY: I'd say I have but I don't now. I mean I would do – there's nothing wrong with that! [we laugh]

BEN: So why is it less important to you now?

CAROL: I don't know. I mean, quite a lot of the time I'd rather just go out and enjoy myself, because if you're on the pull...I don't think girls ever go out and talk to blokes like they talk to us. I'd never go out to pull [right], and I don't think it's like that at a club anyway, because you can talk to hundreds of blokes and there's nothing in it at all [no], whereas if you're going to a pub and a bloke comes up and talks to you or you go over and talk to a bloke they automatically think that you fancy them, which isn't the same in a club where you can just go up and talk to anybody and they don't think anything of it.

While these are clearly significant experiences and are undoubtedly not unique to Carol and Lucy, certain academic accounts of women's relationships to clubbing appear to overstate and, at times, romanticise this process of what might be referred to as 'de-sexualisation' (Ward, 1993)[5]. In this brief examination of the alternate interactional demands of clubbing that some of the clubbers to whom I talked conveyed during their interviews, I want to make use of and attempt to modestly 'texture' some of the work of Maria Pini (1997a; 1997b) on women and their relationships to raving. Pini (1997b) notes how, in her research, women involved in the early British rave scene of the late 1980s and early 1990s appeared to experience the scene as sexually progressive inasmuch as it 'enabled an escape from the traditional associations between dancing, drugged-up women and sexual interaction' (cited in Pini, 1997a: 117). Pini suggests that women in her study all saw raving as a 'movement away from sexual pick-up' and towards dancing, and she proposes that raving's main appeal 'seemed to lie in its ability to both offer [the women] new ways of experiencing themselves, and to transform their under-standings of intersubjectivity' (ibid.: 118). Rave, she suggested, was not so much concerned with 'changing the world' as with the 'constitution of a particular mind/body/spirit/technology assemblage which makes for an alternative experi-ence of the self' (ibid.: 118).

In an earlier but related discussion about women and the 'rave culture' (to which Pini also refers), McRobbie (1994b) similarly proposes that dance culture has opened out a new possibility for women to be confident and prominent in forms of youth leisure. In rave, argues McRobbie, the 'ladishness' is replaced by friendliness, permitting 'pure physical abandon in the company of others without requiring the narrative of sex or romance' (ibid.: 168). Women who went 'raving' experienced the new-found tension that comes 'from remaining in control, and at the same time losing themselves in dance and music' (ibid.: 169).

Yet Pini and McRobbie present only a partial picture. First, as Thornton notes, 'these studies tend to conflate the *feeling* of freedom fostered by the discotheque environment with substantive political rights and freedoms' (1995: 21; emphasis in original). Thornton further notes how the lyrical content of many of the dance tracks, and this is especially so in many of the 'rave' tracks of the early 1990s, helped to promote this conflation – 'the lyrics of dance tracks, which raid the speeches of political figures like Martin Luther King and feature female vocalists singing "I got the power" and "I feel free" work to blur the boundaries between affective and political freedom' (ibid.: 21–2)[6]. Second, while a number of the clubbers, such as Carmel (p. 41) and Carol and Lucy (p. 42), would seem to support both Pini's contention of changing inter-subjectivities and McRobbie's related suggestion that the physical pleasures of raving become, for women, somehow relatively desexualised, other clubbers described qualitatively different experiences of clubbing in which the practices of sexualisation were as – if not more – important as within other social spaces. These clubbers seemed to experience liberation in completely the opposite fashion; through what might be termed '*hyper*-sexualisation'.

SARAH: I'd be lying...okay, I go down there because...I like flirting, and for me it's like my form of safe sex [uh-huh]. If I go down there, I look good, y'know, loads of blokes that are in bands that are famous – they've got screaming women after them – they'll look at me and go 'Oh who's she, she's really nice?' [uh-huh]. You know – come and buy me a drink, and that's as far as it goes, I'm happy, I've had a dose – I can go home. It's the attention thing – it doesn't have to lead to anything [no] and the few times that it has it's always ended up trouble so I'm very wary of going out with anyone connected with that scene 'cos I've been out with two people that have been in bands and they're just complete head-fucks – it just doesn't work, unless you're in that business, totally. For me, I get more of a high, an adrenaline rush, if I end up just hanging out...to me that's just as good sometimes. That's a successful night.

That the 'abandon-without-intent' of which McRobbie speaks (p. 43) is not a consistent feature of clubbing is apparent in this extract from my first conversation with Sarah. Sarah claims to enjoy flirting with men when clubbing, on the one hand, she also enjoys the 'adrenaline rush' of just being with people – 'just hanging out'. When she does flirt, Sarah appears to be more concerned about being *seen with* someone than about being with someone as the prelude to a fully consummated sexual experience. Clubbing for Sarah appears to be as much about sexual display and flirting for *herself* as about being with others. As Sarah puts it, clubbing is her form of 'safe sex'.

The large-scale Release (1997) survey of 520 clubbers presents confusing results in terms of the links between sexual interaction and clubbing. For example, while 45 per cent indicated music, 37 per cent indicated socialising and 35 per cent indicated atmosphere as important to their clubbing experiences, only 6 per cent of clubber respondents suggested that 'meeting prospective sexual partners' was important. Yet Release also notes that of those surveyed, over half had actually had (or claimed to have had) sex with 'someone they had met at a dance event'. This would suggest that while sexual interaction *during* the clubbing experience itself appears *less significant* than one might imagine, clubbing as a meeting place for prospective sexual partners remains important. A further point of note in the Release (1997) study was that men appeared slightly more likely than women to have met sexual partners while clubbing. Thus, while Sheila Henderson (1993) suggests, somewhat incredibly, that during ecstasy (MDMA) experiences 'most men have the opposite to an erection: a shrinking penis' – and one cannot help but marvel at what must have been an ingenious methodology –

the Release (1997) survey suggests that 15 per cent of their male respondents had met two or more sexual partners while clubbing, and 19 per cent had met *three* or more. The disparity between the Release survey results and the individual accounts that Pini, McRobbie and I have gathered in our research on clubbing would suggest that it may be possible that during interviews about clubbing, clubbers, of whatever gender, downplay the sexual facet of clubbing and almost idealise their experiences as emancipatory.

I argue later in 'The night out' that this idealisation of the clubbing experience in no way devalues what are clearly, for many, powerful sensations. However, for the moment what is evident is that, in contrast to the more generalised claims of Pini and McRobbie, some women clubbers experience clubbing as involving a level of often heightened sexualisation or sexual display. Yet, and this is where I believe Pini does get it right, while this sexualisation infers that being seen as sexy is important to these women, it is less apparent that they are presenting their sexuality in this fashion in order to find a 'partner'. It may well be that these women are constructing themselves as 'sexy' for *themselves* because they cannot or do not express their sexualisation in the same relatively open way in other areas of their lives – their work, their homes, and the pubs, bars, restaurants and other social spaces they visit[7]. In this way women may exhibit a confidence and form of almost self-sufficiency – a dis-interest in compulsorily and inevitably consummated sexual interaction – which is unlike that which they might exhibit or experience in, say, a pub. The underlying codes of interaction are quite different, and this difference appears to be attractive to women.

I would therefore partially agree with Pini and McRobbie in their general contentions regarding the apparently liberatory aspect of clubbing for *some* women. However, I would contextualise these suggestions by proposing that it is less that women become de-sexualised, for they do not, just as it is quite obvious that men do not suddenly become disinterested in sex. Rather, their sexuality becomes *differently* important. Some women choose to play-down their sexuality, others choose to exaggerate it. Some women are less concerned with sexual interaction during clubbing than they would be during, say, pubbing, while other women are *more* interested in sexual interaction[8].

For a variety of further reasons, such as the difficulty of verbal communication, clubbers' enjoyment of and focus upon dancing, the lack of spaces within the club for sitting and chatting, and the presence of drugs such as ecstasy (MDMA), which can suppress the libido (Saunders, 1995), clubbing is often constituted more through the practices of dancing within the clubbing crowd than on one-on-one social interaction. Finding a partner will, of course, still be a concern, as the Release (1997) survey rather confusingly infers, but a concern that may take second place behind and beneath a desire to dance and to interact with the clubbing crowd *as a crowd*.

Undoubtedly, the codes and demands of interaction in clubbing differ from those in many other social spaces. It would appear that for many woman it is this very difference between the overtly predatory interactional ordering of, for example, pubs, and the more 'look but don't touch' codes of sexual expression in clubbing that contributes towards the attraction of clubbing. Clubbing constitutes a different experience of togetherness than that experienced in other social spaces: an experience less explicitly centred on couples (and 'one-on-one') and more within notions of being within the crowd ('one-among-many'). The flux between the experience of identities, including sexual identities, and identifications that the practices and spacings of the clubbing crowd can facilitate appears critical.

Belongings and identifications

The logic of identity is, for good or ill, finished.

(Hall, 1991: 43)

The twin notions that clubbing as a form of social space is qualitatively different from the 'city streets' beyond and that clubbing involves alternate orderings, codes and modes of social interaction are linked. The common thread that cuts through both of these conceptions is that of *belongings* and of the search for spaces and experiences of identifications or affective gatherings (Maffesoli, 1995).

BEN: What do you enjoy about clubbing...I mean particularly enjoy?

SUN: I could come up with...I'd need to think for a couple of seconds [um-huh]... ...basically ummm...I've grown up believing and being told that life is pretty grim and you've got to fucking work your bollocks off 'cos otherwise you're not going to get anywhere – I've still a very strong work ethic, that's inherited from my Asian background...ummm...but what I really like about going clubbing is...is...just, it's just the trash Western, hedonistic nonsense of it all, y'know, just going out to have a good time for the sake of it, and just meeting people and actually being there on the scene – I remember in the 80s sitting around and being depressed because we'd missed punk y'know? [uh-huh]. And this is happening now and I just think this is my culture, this is the only thing that I have that I can call home, I'm not English in the normal sense of the word, English, I'm not Indian...by any means, ummmm...I've just been dumped into this urban nightmare scenario with a blank sheet of paper [right] and I just, I just love it – it's my culture, it's my home, umm...to me that's the nearest I have to a...well, it is, it's my culture, it's how I choose to spend my time ummm...I just think it's...Julie Burchill once said 'Western civilisation is the only culture where life isn't a toss up

46

between horror and boredom' and I really like that. I like Western twentieth century flash, tacky materialist culture because it's just so fast and so quick and always moving and it just keeps me, my head, active [ummm] and it gives me something to do and I really enjoy it [umm]. Y'know it's... I think actually that sheer hedonism is actually in a way more profound than y'know poetry and literature and art and philosophy and mathematics and all that stuff [right]. To me this is the fucking point, this is the life, y'know? [ummm] I mean, my parents came over from India because, y'know, despite them not knowing anything about the West they knew that people had a better life there, they did have a fucking better life, y'know – in India, people my age are just working their butts off and no joy! And they're aiming for something and this is it [umm]. And I figure, I'm young, I'm rich, I'm in the middle of London when it's really happening – I mean jungle is just suddenly...ramped up the notches so high [umm]...and fuck it, I'm going to do it! Yeah! And when I'm 30 maybe I'll settle down and have kids and at least I'll have some-thing to tell them, y'know, I was there...[9]

This passionately nuanced story raises a plethora of issues. I draw out just five related points. First, for Sun clubbing is a form of trashy hedonism that he equates with the West. Second, this trashy hedonism – just doing something for the sake of it – is what Sun believes life is, or should be, about ('just going out to have a good time for the sake of it') – it is what he wants to do. Third, Sun specifically highlights a 'youth subculture' from the 1970s and 1980s which he missed – punk – picking out meeting other people and 'just being there on the scene' as experi-ences that went with it as seeming valuable. Fourth, he therefore appears to understand jungle music as *his* culture in a similar way that punk was for some of those before him. This search for a culture with which to feel belongings appears to stem both from Sun's awareness of the nature of previous 'youth cultures', such as punk, but also from his particular biography, involving 'Asian' roots and his subsequent search for a 'home': 'I'm not English in the normal sense of the word, English, I'm not Indian...by any means'. Fifth, although clearly passionate about jungle music and the senses of identification that he feels the jungle culture can provide, Sun is also already looking forward to the future. He wants to be able to say 'I was there' to his children; to evoke his youthful belongings; to show how and where 'he belonged'.

This notion of belongings – of a quest for a sense of 'home' – which jumps out so forcefully from Sun's story, is, I propose, at the very heart of the clubbing experience for many clubbers. Of course, each will narrate their particular understanding of what this belonging means to them in different ways. In 'The beginnings' I introduced debates concerning the supposed changing nature of our social relationships, both with ourselves and our own identities and also with others and social groupings more broadly. I now re-visit and deepen these points specifically in relation to clubbing in suggesting that clubbing is partially constituted through this search for belongings.

Identity and identifications

It is relatively widely accepted that, although still partially effective, for example, politically and economically, the pre-eminence of the logic of identity as *the* structuring feature of society has been shattered by the increasing cultural fragmentation (Featherstone, 1995) of the world. This fragmentation is what Marcus (1992: 311) refers to as 'global creolization processes'. Increasingly, 'micro-structurings' of shared sentiments, experiences and emotions appear to be playing a more important part in people's everyday experiences, particularly in relation to the social groupings and affiliations through which individuals constitute themselves (Maffesoli, 1995).

Stuart Hall (1991: 44) describes these historical transformations that have disturbed the 'settled logic of identity' as 'theoretical decentrings', giving as examples the instability of the nation-state, the changing fortunes of nations' economies, and the consequent challenges to notions of nationhood and national identity, as well as to the 'fragmentation and erosion of collective social identity'. By the latter, Hall means the 'large-scale, all-encompassing, homogenous...great collective social identities of class, of race, of nation, of gender, and of the West' (ibid.: 44). Although, as Hall contends, these great collective social identities have not disappeared altogether, their disrupted location within our conceptions of the world suggests that they cannot be thought of in the 'same homogenous form' (ibid.: 45). In short, these *ascribed* collective identities 'do not give us the code of identity as...they did in the past' (ibid.: 45–6)[10]. New formations of identities and identifications have taken their place; new formations that prompt Hall into making a number of crucial points about the changing nature and role of identities in contemporary society. First, Hall suggests, it is clear from the 'new theoretical spaces' which are opened up by these changes that identities are never completed, never finished – 'identity is always in the process of formation' (ibid.: 47) and our sense of identity 'can never be resolved' (Chambers, 1994: 25). Second, identity 'means, or connotes, the process of *identification*, of saying that this here is the same as that, or we are the same together' (ibid.: 47, my emphasis). Third, this notion of identification is always constructed through ambivalence; that is, it is always constructed in terms of belongings and outsiders (Chambers, 1994), between 'those who belong and those who do not' (Hall, 1991: 48).

On a more everyday, experiential level, the notion of identifications that Hall presents here highlights the empathy that we can, though do not all, nor always, feel with others when we are in a group situation (Maffesoli, 1995; Hall, 1991; 1992). The *process* of identification is one in which 'people come to feel that some other human beings are much "the same" as they are and still others are more "unlike" them' (de Swaan, 1995: 25). Certain social situations – and I suggest that clubbing may be one example – foster a going-beyond of individual identities, an experience of being both within yet in some way outside of oneself at once. In extreme circumstances this experience of ambivalence through identification can be referred to as a sensation of ecstasy or *exstasis*[11]. While proximity to others is not a necessary pre-condition, physical co-presence is often actively sought in the establishment of identifications. This is a search for togetherness that Boden and Moltoch (1994: 258) refer to as 'the compulsion of proximity' – the need to be near others who are understood as broadly similar to oneself.

BEN: How does clubbing fit into your overall lifestyle at the moment?
LUKE: I hate to admit it, but clubbing gives you sort of an identity. You belong to a relative small group of people, with very different interests and skills, but there is something that connects you. I like being in a city because it's quite anonymous, but it feels very good to find steadying points in the city. Clubbing makes me feel at home in a city, because you meet lots of people. First in the club, and later you'll meet them in the streets occasionally. I've been into other scenes for a short time as well...a bit of skating and grunge...but I always get bored with them. House seems to be the only thing that comes back. Maybe it's very easy to get into. Compared to my 'adventures' in other scenes, it seems to be very easy to get into a club scene. In less time and with less effort, I meet more people there. I don't know whether that says something about the scene or more about myself.

While he 'hates to admit it', Luke does see clubbing as providing him with 'an identity' – 'you belong to a relative small group of people, with very different interests and skills, but there is something that connects you'. This is a great evocation of the notion of identification. Luke then builds on this evocation by expressly linking it to his experiences both of living in the city generally – clubbing makes him feel 'at home' in the city – and to his own history of other 'scenes' such as 'skating and grunge'.

Following this story of Luke's, then, identifications might be understood as *elective forms of identity* (Scheler, 1954). They are 'the rendering to someone of identity' (Friedman, 1992: 332). Identifications are affectual in nature, usually small in scale, and might be premised upon an affiliation with a charismatic leader, a totemic or tribal symbol and/or a shared emotional experience (for example, a shared appreciation of music) and a shared site (Hetherington, 1996).

The importance of tactility and proximity – of a sharing of a bounded space – to most identifications restricts their size, relative to slightly less tight concepts and notions of 'lifestyles', to a smaller scale, and this usually means a heightened prominence for the practices of face-to-face interaction (Chaney, 1996; de Swaan, 1995). Through participation within an identification, individuals may define themselves according to the others who they see as also constituting that identification. Identifications can, in this sense, be constructed as virtual 'mirrors of identity' – you see what you are or what you want to think you are (Goffman, 1963).

Identifications do not replace identities. An identification, once established, 'does not obliterate difference' (Hall 1996: 3); identity retains a powerful significance. What is being proposed is that at certain times and in certain places aspects of our identity – or elements that constitute it – are submerged beneath a usually transitory feeling of social identification. In these situations, notions that are central to our *personal* biographies – our understandings of our own and others' gender, ethnicity, social class – can become temporarily eclipsed by what it is that we share with those with whom we are co-present. First and foremost this is usually our occupation of (a) space. More important than what divides or what distinguishes us as individuals is what unites us or what we share, even if this is only the space and time that we temporarily co-inhabit. The submerged aspects of our identity – our personal biographies – may still be resonant, yet in a muted, almost subconscious fashion. This construction of identifications thus challenges the predominant sociological focus upon the individual without dismissing it. The contexts and forms of our focused group or social interactions become central to our sense of our selves (Goffman, 1959). In contradistinction to some notions of self-identity, in which increasing informalisation and individualisation of consumption practices are deemed to produce a weakening uniformity of consumption choices and experiences, and in which the constant amendment of the self is achieved through a 'parading of its "identity"' (Warde, 1994a: 882), identifications cast the practices of *consuming* as being both generated by, yet also contributing towards, the sharing of styles, of spaces and even of emotions and sentiments.

Sun and Luke's stories about their desires for a sense of belonging and identification through clubbing are examples of the shared emotional understandings that clubbing can provide. An understanding of the practical constitution of these identifications is deepened in the following stages of 'The night out'. However, establishing these identifications is about much more than merely displaying the correct clothing or demonstrating a knowledge of the musical forms involved in a clubbing genre. Issues of distinction and stylisation are more complex than this, and it is to these practices of being 'cool' and demonstrating 'style' that I now turn.

CLUBBING AND COOLNESS: DISTINCTIONS AND BELONGINGS

The sensations of belonging that appear so central to clubbing are partly constituted through the processes, practices and experience of being cool (or not) while clubbing. Styles of dress, of speech, of dancing, choices about which beer to drink, about clubbing venues, or about drugs may all be central to the notion of belongings through which clubbers can experience clubbing as part of a collectivity (a crowd) as opposed to individually (or atomistically).

When sharing a style, which, in one sense, might be understood as a more everyday way of describing a facet of the sociality that I introduced earlier, people can – though do not always – identify themselves as a group, as having something in common. What is shared may temporarily become more significant than what is distinct. Further, this sense of identification may also be perceived from the 'outside' – the group may be identifiable as such by those that are not, whether through choice or exclusion, part of that group.

BEN: Tell me more about this word 'trendy' that you use?

SIMON: Ummm... ...well... ...clothes, movements, the way that people look at you, ummm...

BEN: What do you mean?

SIMON: Umm, trendy was Scott's word firstly, secondly umm, it may not be whether people are trendy whatever that is, as in they're following a trend or they're setting a trend of whatever but more importantly whether they think they're trendy [yeah] and I think that's something that you can see in people's eyes, especially women because I'm more used to...well there's definitely a difference in the way that men and women look at each other in social circumstances like bars and clubs.

BEN: So, to carry on using the word for a minute, it was a trendy night...or...

SIMON: Yeah that's why people feel justified in looking at you and you get the sort of raised eyebrow or they just check you out. They check you out to see if you're as cool as them. It's a process that I don't really go through but this is what I've picked up from other people looking at me. In places – not all the time, this is going to sound ridiculous – but I actually I find it more in pubs in Soho at 6 o'clock on a Friday evening when everyone's sort of drinking their heads off and all the fucking media types from Soho are there checking each other out and you come along, a fucking yokel from the sticks and order a couple of pints of cider and they look across at you with their cappuccinos and gin and tonics and you wonder what they are thinking. I don't think that this is entirely in my head, I actually think that some people in London are quite good at giving you these looks. It's self-defence for them maybe but erm...

BEN: So you can feel them looking you up and down?

SIMON: I feel that 'checking you out' is a better phrase.

BEN: Do you do it?

SIMON: No, not as much. I really don't think that I do. It's all suddenly turned into something that sounds ridiculously paranoid fantasy, but ummm, I suppose I'm definitely...I don't want people to think that I'm a fucking joke when I go out in any social thing and so I'm always slightly worried that people are thinking like that, and ummm, because London is such a ridiculously ambitious, social-climbing, money-grabbing, tight-arse place people are good at checking each other out in this way. It's just a kind of showing their feathers; it's posture. You know 'are you a stranger...are you some kind of big mover and shaker or are you a punter...are you likely to come over and fucking elbow someone in the face?'. I mean these are all questions that go through your mind, although less so last night. I think what you said about feeling comfortable in the crowd, I can't deny that I felt that I wasn't in any way in any physical danger at all in that club last night. It seems odd to bring it up but there have been times in the past when I've been in clubs and I've been worried about that slightly. There are people in clubs who will go up to you and try and get you to go to an after-party party and they'll tell you where it is and it will be in some fucking no-go area of London [uh-huh], some really rough place, and you'll think 'oh that sounds cool, I'll check that out' and then you get there and you realise that someone's been pulling your leg.

It is immediately apparent from this story that Simon spends a considerable amount of time thinking about how he and others in the clubbing crowd are presenting themselves to each other as actually part of, as well as before and subsequent to, their clubbing experiences. Of particular significance is Simon's notion that it is less important whether you actually *are* cool than whether you *think* you are cool. This would seem to confirm the suggestion that 'coolness', for some, is about more than the display of certain clothing or techniques.

In experiencing oneself as cool (or not) and judging others as cool (or not), complex processes of 'distinction' (Bourdieu, 1984) and differentiation are involved, as well as learned or acquired practices and techniques of 'belonging'. In this second part of the section, I look in detail at the notion of coolness and discuss its centrality to understandings of belongings in the clubbing experience.

I make four brief points. First, I briefly discuss the work of Pierre Bourdieu (1984) on 'distinction' – his sociology of taste has recently been used by Thornton (1995) in her examination of the creation of clubbing genre and 'taste cultures' through various media. Second, I attempt to go beyond or 'behind' Bourdieu's notion of distinction and the establishment of difference in suggesting that the clubber chooses the clubbing crowd as much as, and probably before, the stylisation which may characterise that crowd, so in the clubbing crowd there are as many concerns with stylisation and establishing affiliations as there are with establishing differences. Third, I contextualise these points through addressing the different conceptions of what is cool and the hierarchies through which these conceptions are constituted, particularly in respect of notions of the 'mainstream'. Fourth, I attempt to bring these first three points together in suggesting that choice of club night becomes a vital moment of self-identification and is thus entangled in processes of identity formation and amendment.

Cool and distinctions

> Consumption is...a stage in a process of communication, that is, an act of deciphering, decoding, which presupposes practical or explicit mastery of a cipher or code...A beholder who lacks the specific code feels lost in a chaos of sounds and rhythms, colours and lines, without rhyme or reason.
>
> (Bourdieu, 1984: 2)

In his analysis of the ways in which various consumer goods and the form of their consuming are used by specific groups in the process of demarcating their distinctive way of living as being different from that of 'others', Pierre Bourdieu (1984) sets out how different status groups in the French society of the 1960s and 1970s had differing access to various forms of capital. Bourdieu suggests that through this 'distinction' they distinguish themselves from those in other status groups.

Access to *economic* capital – money, wealth, real estate, factories, shares – provides a different form of distinction from those with access to *cultural* or intellectual capital, which is the ability to be able to talk and write about arts, culture, music, and to create new cultural products through this skill. It also provides a different form of distinction from those with *social* capital – who you know, and who knows you (Bourdieu, 1984). For Bourdieu, cultural distinctions are used to support class distinctions, with notions of taste – as an ideological category – functioning partly as markers of 'class', where 'class' is understood both as a socio-economic categorisation and as a particular level of quality (Bocock, 1993).

Bourdieu attempts to fuse a conception of socio-economic structuring in capitalist societies with that of the structurings of cultural symbols and signs. He highlights the role of education in reproducing over time the structurings of society in cultural (symbolic) and socio-economic terms. In this way Bourdieu establishes a link between consuming *practices* and symbolic or imaginary cultural *meanings* (cultural capital). Thus, consumption is not merely, or not only, about the satisfaction of biological needs, such as eating and staying warm, but is also about the creation of, and affiliations with, status groups. Acquired tastes for specific sporting, musical or other aesthetic forms are cultivated and developed into 'good' tastes through both formal and informal educational settings, as well as within the status groups themselves. Differences in taste are used in this way as forms of distinction[12].

Building upon Bourdieu's notions of taste and cultural capital, Sarah Thornton (1995) opens out an understanding of the ways in which coolness, or what she refers to as 'hipness' (ibid.: 11), works in clubbing cultures. Thornton conceives of coolness as being a form of subcultural capital:

> Subcultural capital confers status on its owner in the eyes of the relevant beholder. In many ways it affects the standing of the young like its adult equivalent. Subcultural capital can be *objectified* or *embodied*. Just as books and paintings display cultural capital in the family home, so subcultural capital is objectified in the form of fashionable haircuts and well-assembled record collections...Just as cultural capital is personified in 'good' manners and urbane conversation, so subcultural capital is embodied in the form of being 'in the know', using (but not over-using) current slang and looking as if you were born to perform the latest dance styles.
> (Thornton, 1995: 11–12; emphasis in original)

This is an effective approach to developing understandings of the ways in which clubbers define, locate and affiliate themselves with larger social formations within wider clubbing cultures. In this way Thornton infers that deficiencies in the subcultural capital required to decipher the 'codes' of clubbing can lead to exclusion through the 'condescension' and derision of those that have acquired the necessary codes (Bourdieu, 1984: 57). Thornton goes on to discuss in detail the distinctions that characterised early 1990s clubbing in terms of media constructions of 'good' and 'bad' taste, and in particular focuses on the distinction between the mainstream and the 'hip'. For Thornton (1995: 14; emphasis in original) 'the difference between being *in* or *out* of fashion, high or low in subcultural capital, correlates in complex ways with degrees of media coverage, creation and exposure'. I want to branch out on a tangent and consider the crowd context of these processes of distinction. Thus, turning more to the identifications produced through these distinctions themselves, I flip these processes of distinction over and look instead at the relationships between stylisations or shared notions of what is cool, and belongings.

Cool and belongings

Simon's story about trendiness, coolness and 'checking people out', presented at the beginning of this section, is indicative of the intensive monitoring of self and others that is an important part of the clubbing experience, and, indeed, of social life more generally. Clubbers distinguish themselves from others through their tastes in clothing, music, dancing techniques, clubbing genre and so on. These tastes are trained and refined, and constantly monitored not only in order to distinguish oneself from another, but also in identifying with those that share one's distinctive styles and preferences – two subtly different aspects of the same processes.

DWAYNE: I'm very elitist, I think you have to put a lot of effort into training your tastes, go to a lot of places to do it, and a lot of people for various reasons – because they don't have the time, because they don't have the resources – just don't do it. Maybe they just don't want to. They cannot understand these different styles. For them, they get it later, when it's mainstream. I'm very elitist because of my experience…it has to do with the way I perceive taste works. It works strangely. There are social connotations of musical styles and so…if people I didn't like liked the music I like, it would spoil the music. And as I said before I think that there's a very close connection between the music…how one likes music, and the kind of people one conceives of as haters or fans of that music [yeah]…

BEN: So…do you like the people around you to be LIKE you?

DWAYNE: No, of course not, I want them to be much COOLER than myself. I want to feel 'oh God, this is really exciting, this is so brilliant'…I feel uncomfortable if I'm surrounded by 15 year olds, but um…it's a question of body language, dress and probably a lot of what I fantasise into these people, which is probably, in many cases, totally away from what they're really like [umm]…but, I don't talk to them most of the time so I have no way of finding out.

Dwayne openly admits to being 'elitist', to having 'trained his tastes' – a process that he appears to see as inevitable and, he infers, almost obligatory. This point appears to offer support to the Bourdieuan thesis of distinction, as well as to Thornton's theorisation of 'subcultural capital'. Dwayne clearly differentiates himself from other clubbers who 'because they don't have time, perhaps because they don't have the resources…cannot understand these different styles', but who, instead, 'get it later, when it's mainstream'. However, in addition to these important points about differentiation from others is a further set of points about Dwayne's relationship with the crowd around him and his enjoyment of being

within that crowd. He holds strong feelings about with whom he desires to identify – feelings so strong that they can result in Dwayne being put off certain music
that he would otherwise enjoy purely because clubbers he perceives as less cool
than himself appear to like that music: 'if people I didn't like liked the music I
like, it would spoil the music'. When Dwayne goes clubbing he wants to find
himself – in both senses of the term – among other clubbers whom he perceives
as being *much cooler* than himself. Dwayne wants to identify, and be identified
with, clubbers who move, dress and occasionally talk in ways that are cooler than
how he perceives himself – 'body language, dress and probably a lot of what I
fantasise into these people' are all important.

Yet, despite actively wanting to identify with those whom he perceives as
cooler than he is, Dwayne openly admits that he has no way of knowing whether
they are actually cooler than he believes himself to be because he does not talk to
them. Vitally, this appears not to matter to Dwayne, although it is, again, indicative of the alternate interactional orderings that can characterise clubbing. As
Simon suggested earlier, in one sense it is less important whether people *are* cool
than whether they *think* they are cool. In a similar way, it appears that, for
Dwayne at least, it is less important whether those around him are actually cooler
than him than they are *identified* by him as being cooler than himself. This leads to
the possibility that the clubbers whom Dwayne thinks are not cool may well
understand themselves as being cool, and may even perceive Dwayne as *un-cool!*
Thus, coolness is not just a quality that is ascribed and seen to be possessed, but
also a hierarchically based and contextual sensation that is produced through
mental–physical processes of striving, desire, fantasy and imagination.

The crucial point here is that *other clubbers* within the clubbing crowd of which
he is a constitutive member are central to Dwayne's understandings of both what
is cool and what is not, as well as to his construction of himself and his identifications with that crowd. Thus, processes of distinction and questions of taste
constructed in advance certainly run through the constitution of the night for
Dwayne. Yet, so also do imaginings and fantasies – his fantasies – about the coolness of those with whom he might be associating – and thus how he in turn *might*
be perceived by others – and about his belongings within that clubbing crowd.
Identifications and concerns about stylisation and belonging might thus be understood as complementary and equally important processes to those of distinction.
These notions of belonging clearly foreground the proximity and tactile quality of
being within the clubbing crowd itself.

Clubbing groups and the social formations constituting club nights might thus
be better understood in similar terms to Maffesoli's (1988b; 1995) 'neo-tribes':
micro-groupings that develop as a result of the constant comings and goings
between social situations that occur in society and which are premised upon a
superseding or submerging of identities (Maffesoli, 1988b). For Maffesoli, these

'tribal formations' foreground the immediate and the affectual and are marked more by their practices and spacings of inclusion and exclusion and their shared styles than by more formalised memberships and contractual codes of behaviour – the latter being reminiscent of notions of lifestyle (Shields, 1992c). Although I have certain reservations about the supposed newness (the 'neo-') of these social groupings, the conception of 'neo-tribes' outlined by Maffesoli does appear to recognise the centrality of, for example, the sense of emotional community, the ephemerality, the importance of atmosphere, and the key notion of emergent identifications that characterise the belongings of clubbing in contrast to more rigid notions of lifestyle. Maffesoli (1995) differentiates these relatively unstable and multiple 'tribal formations' from those associated with notions of more conventional 'lifestyles' by proposing that it is less membership *per se* that is important than the fluid movement *between* different temporary memberships. Thus, 'neo-tribes' prioritise the here and now, the affectual and the tactile[13].

In short, the processes of identification that I have been outlining are suggestive of how, in clubbing, belongings can come before and at times supersede identities (Warde, 1994b). That is, clubbers seek to identify with others not just through the physical display of the insignia of clubbing, for group identification requires subtly more than the accoutrements of specialised language, certain techniques of interaction, shared tastes and familiarity (Warde, 1994b). The belongings constituted through shared membership of the clubbing crowd can *only* be experienced in that form and to that intensity *in the clubbing crowd* in which a sharing of space is possible, and socially so powerful. The individual clubbers' imaginings and constructions of their own styles thus take on their most powerful meanings in the presence of others whom they perceive to be similar to, or even cooler than, how they perceive themselves.

Given this prominence for co-presence, tactility and the affectual in the development of identifications, it follows that the perceptions and practices of belonging are both reflexive and embodied. They are constituted through the languages of the body, such as gesturings, and accepted practices, such as dancing techniques, as well as through elements of dress and fashion, and are premised upon a clubber's understanding of what is cool and what is not cool at varying times and in differing situations. Furthermore, the nature and extent of these perceptions and practices also depend upon whether a clubber feels that *they* themselves are cool, and if they do, how cool.

The embodied practices of 'coolness'

Understood one way, the management of cool in the achievement of belongings can be approached through the inversion of Erving Goffman's (1968) study of *Stigma*. In *Stigma*, Goffman looks at the management of tension for the discredited, and of information control for the discreditable. I am looking at the notion of style as used and practised by clubbers and am interested in the management of attention for the stylish and those who perceive themselves as cool, and the management of information control for the potentially stylish – those who aspire to be seen or understood as being cool. Goffman looks at those individuals who are in possession of 'attributes which may be deeply discrediting'. I am looking at a situation where individuals may be, or may aspire to be, in possession of attributes that may be socially desirable or attractive. Where Goffman concentrates upon the management of a 'spoiled identity', I am instead interested in the management of what might be the opposite – a 'cool identity'. Where Goffman is explicitly concerned with notions of stigma as expressed or controlled (suppressed), I am interested rather more in expressions of coolness. Apart from an explicit recognition of the multiple hierarchies that operate within notions of coolness – what is cool to one may be un-cool to another – what this Goffmanian reading of 'coolness' adds to my unfolding conception, above all, is an emphasis upon on-going, contextual and embodied practices. Belongings are about much more than choosing the 'right' clothes, although this can also be important. Belongings are also practised through the techniques of sociality on a more embodied level and with an awareness to the continually changing social demands of the encounter.

Cool and COOL

It follows from earlier that while group belongings can be established through shared conceptions of what is cool, then notions of not belonging can also arise through a disparity between these conceptions, as Dwayne's story demonstrates. There is no fixed notion of what is cool and what is not, for not only will each individual clubber understand coolness in different ways, but, within the identifications and hierarchies of belonging that notions of style partially constitute, there will always be sub-hierarchies and sub-sub-hierarchies. Where one clubber sees cool, another will see non-cool.

What is cool (or 'hip') is frequently contrasted with 'the mainstream' (see Thornton, 1995[14]). The actual meaning of the word 'mainstream' is difficult to ascertain and few people appear to define themselves as 'mainstream' clubbers – I have yet to meet *anyone* who has done so – yet an awareness of what the term appears to represent is widespread. The 'mainstream' in clubbing is usually used to refer to what the clubber is not, whatever their preferences and resources of subcultural capital (Thornton, 1995).

MARK: I'm trying to make a structured argument here!!! [we laugh]. Basically you have, I mean it's quite easy, basically you have your like your out of town suburban pick-up joint type of place [okay], which are your, like where I live, The Regal, and…then you've got like your normal clubs in central London which attract people from all over this place, as opposed to just the suburb where they're in [right] say, Pleasure Lounge or something like that…Now, what you've got in London which you haven't got in most other places – you've got the super-club phenomenon where you've got your Temple, your Club GB arguably, things like that, places like that. They're on every tourist's map [I laugh] – some people adore them, some people hate them, I mean I've got a friend who goes down to the Temple every single week. They absolutely love them. Now you see I would absolutely…I've never been there, actually, but…let's put it like this, every time I say I want to go there, I want to experience it so I can write it off, okay [right], my friends who are going to take me go 'oh I had such a bad time last week, let's not bother, okay?', and it's like that kind of club phenomenon where you have VIP rooms and things like that, which just doesn't interest me at all. Then, which leads me on to the places like what I call 'underground clubs' which is like most of the places I go to, not Paradoxical YOUniversity – most of the places I go to, ummm…where, there's no security, there's no dress code, you're going to get in whatever you're wearing basically [okay] and it's cheap and it's all on a fairly communal basis really…

Mark draws out a number of distinctions between his preferences in clubbing and 'other' forms or genres of clubbing. First, he suggests that suburban clubbing is, for him, qualitatively different to clubbing in central London in terms of the constitution of the crowd. Second, within central London he distinguishes between 'super-clubs' – clubs that 'are on every tourist's map' – and 'underground clubs', in which, and contrasting with the former, 'there's no security, there's no dress code, you're going to get in whatever you're wearing…it's cheap and it's all on a fairly communal basis'. Mark's preferences are clearly contrasted against and defined in terms of his perception of other clubs that he sees as uncool, despite having 'never actually been' to these other clubs. Thus, for Mark, the 'super-clubs' such as the Temple – which other clubbers whom I interviewed, such as John and Sun, saw as cool – are the equivalent of the 'mainstream'.

The 'mainstream' plays the role of 'the other' for clubbers. The 'mainstream' is everything that they stand against; everything that their choice of clubbing experience is *not* about (Thornton, 1995). The description, 'mainstream', is applied to club nights, clothing, genres of music and attitudes that are, in some way, the antithesis of what clubbers perceive as clubbing's *raison d'être* (at least as far as they are concerned). By inference, this is also the antithesis of how they see *themselves*, at least while clubbing. There are many scenes and musical genres of club nights in London, yet the one that most clubbers see themselves as outside, and even in opposition to, are those constituted by the chains of Mecca and Ritzy discotheques, of which The Regal, which Mark discusses, is one example, and Zens, which Kim and Valerie mention, is another. These are 'the site[s] that [have] been consistently identified as the location of the mainstream by dance crowds and academics alike' (Thornton, 1995: 106).

As an attitude, against which other attitudes can be constructed and identified with, the 'mainstream' plays an important role in the construction of some semblance of exclusivity and a feeling of belonging, whatever the clubbing genre concerned. Different clubbers voice their concerns over style, what they see as cool and the presentation of themselves as cool, in different ways – for some these concerns are foregrounded, for others they appear unimportant. The selection of a particular night out from the huge number on offer provides an apposite and timely example of these dual processes of distinction and identification.

Cool crowds – selecting a night

Choice of club night is potentially confusing, given the huge and growing number of club nights on offer. Yet somehow individual clubbers select a night or number of nights week after week. The choice of clubbing experience is tied in with a set of issues about how one sees oneself in relation to the clubbing crowd with which one wishes to identify. Thus, the processes through which clubbers whittle down the vast set of choices to a number of selections, often returned to time after time, provide insights into the construction and handling of the identifications that can be achieved through clubbing. The complex and volatile hierarchies of cool that I have started to sketch out are most obviously in operation in terms of decisions about where to go and with whom. Nights appear to be selected as much in terms of where and with whom the clubbers wish to be seen and be associated, as what the club nights will consist of in terms of music and DJs, as Dwayne's story (p. 55) exemplified. The belongings achievable through clubbing thus pervade even processes of pre-clubbing choice. In many cases the belongings of clubbing are in this way being forged well before the clubber gets into the clubbing crowd, and are merely reinforced and consolidated once entry has been achieved.

Clubbers will have at their disposal a multitude of available sources that may assist them in their choices. The weekly London listings magazine *Time Out*, for example, breaks down the vast choice of listed club nights into various musically based genres in an apparent attempt to ease the selection procedures of the clubbers. Most commonly, clubbers identify various nights and clubs as potentially suitable for them through a mixture of media – listings such as *Time Out*, flyers and, to a lesser extent, fly-posters and, very occasionally, television. Carrying most weight are the recommendations of friends and, of course, previous personal experiences. While listings sources are used by many clubbers in their club selection process, not all clubbers use listings in a conventional manner with the intended purpose of the listing occasionally being subverted as clubbers choose *not* to go to a night that they might have attended purely because it *is* listed.

BEN: Yeah...so how do you find out about these nights then?

MARK: Ummm, typically either from friends who hear from other people who are organising it or from flyers for particular events.

BEN: And where do you pick the flyers up?

MARK: Usually on the way out of events, because as you know, basically as you go out of an event you'll get flyers for similar events [yeah] and it all works like that really. Sometimes there's other parties and you don't get to hear about them and someone will phone you up and say 'Hey! don't go there, come to this one instead, it'll be a lot better' and it usually is. So...a lot of the things that I go to are not one-offs, but are irregular events, I mean things...there's one night you might have heard of called Polypresence [yeah I've heard of it] and that's changed its venue...I think it's been in three different venues so...and it's really quite small. There's another night called Eutopia [oh yeah] which takes place at Thyssen Street [right], but they just seem to do it when they like it and I like things like that.

BEN: How else do you get to know about these parties?

MARK: Flyers mostly [really?], 'cos I put things on my web-site and I get those mainly through flyers, so...and that accounts for maybe 90 per cent of the parties that I'm aiming at [ummm], and the other ones are either semi-private parties, 'cos there's all these type of semi-private parties where record labels put on their own parties and just invite their friends, effectively [right]...things like that I usually find out about, but I usually don't bother going as I wouldn't know enough people [right].

BEN: What about nights like Sahara or Paradoxical? Is that *Time Out* or flyers or...?

MARK: It's funny you should say *Time Out* – if things are in *Time Out* then I usually won't go to them [really?]. I know...it's really changed actually. *Back to the Origins* is probably the best example because I really liked *Back to the Origins* – I still do, I mean, to be honest – but since it's been put in *Time Out* it's basically become really popular. It's on every tourist's, every young clubbing tourist's map of London, okay. It's like, if it appears in *Time Out* I won't go. It's on tomorrow, for example, and I'll go into a newsagents and check it out and if it's there I won't go, and if it isn't I'll consider it...

Some of the subtle distinctions that are involved in choosing between different nights and the relatively complex construction of Mark's notion of 'coolness' are evident in this extract. Immediately apparent is Mark's inferred distinction between 'parties' and club nights. He repeatedly refers to the nights he likes as 'parties', which suggests their relative lack of organisation (and often a licence as well), compared to the club nights, which feature in *Time Out* and are accessible to all. Second, Mark uses *Time Out* to decide what NOT to go to, and thus he relies on flyers and the Internet (he has his own web-site) for information about where to go. Third, friends are important – Mark will listen to friends' recommendations about club nights and the presence of friends at the parties that Mark attends is important enough for him to not go to a party that sounds as though it might be good if he finds that his friends won't be there. Finally, Mark again refers to 'every tourist's map' as being something on which his choice of clubs would not appear. He would appear to group 'tourists' firmly within his non-cool category!

The majority of club nights that are organised to the extent that they appear in listings sources, such as *Time Out*, also produce and distribute flyers to other clubs and, more usually, specifically selected youth-based clothing and music shops, bars and restaurants. Club selection can also be based upon information disseminated through specially produced 'guides', clubbing magazines, Internet mailing and discussion lists and local radio advertisements[15]. Finally, as the stories of Dwayne, Simon and Sarah suggest, the likely constitution of the clubbing crowd at any given night is often an important factor in any decision. Attendance is premised at least partly in terms of a process of self-identification – 'are the others present likely to be like, or cooler than, me?'. While the numerous guides and clubbing listings can provide an indication of what the clubbing crowd might be like at a club night, it is not until the clubber gains physical access to the club night that they find out what the crowd is like and whether they feel that they belong there. However, the negotiation of entry to the club and the experience of queuing can provide an early indication of the general style of the clubbing crowd – what people are wearing, whether there are dress restrictions of a set 'dress code', the broad age, gender and ethnic constitution of the crowd. The door can also provide the clubber with the first real test of their claim to belong.

GETTING INTO THE CLUB

It is not possible to imagine a society that does not make extensive use of ground rules.

(Goffman, 1971: xiv)

EXTRACT FROM CLUBBING JOURNAL (08.03.96): What a queue!! A depressing sight. I'd say that there were about 200 plus a totally separate queue for the guest list. The club is fenced off in an old warehouse and the queue snakes along the fence. We decided to stick it, even though it was quite late [midnight] and drizzling. There were no other clubs within walking distance. About four or five coaches of 'glammed-up' clubbers arrived and they were fast-tracked straight to the front while we just carried on queuing – booked ahead I guess. We were envious. A few girls in our group were busy stuffing cling-film-wrapped Es down their bras as subtly as was possible while one of John's friends was quickly hollowing out a cigarette and using that for pill-storage. We were all a little twitchy, pensive and edgy and this was undoubtedly due to the imminent negotiation of the door – people seemed worried about getting in with their drugs, if they had them, and those that didn't were nervous about getting in anyway. John had on trainers and we all suddenly panicked a little when we saw a man walking away from the front of the queue who also had on trainers – was it the trainers that he was being rejected for? Was it something else? What would we do if John got turned away? The door is one of the few places that clubbers sometimes have to literally prove their identity both in the simple sense of age but also in the more complex sense of passing themselves off as the 'right type of clubber' for that club – proving their 'social identity'. The door becomes a defining moment, the moment where you both partially take on and are partially ascribed an identification.

We arrived at the front of the queue about an hour after we had started queuing, say near 1 a.m., after many nervous cigarettes and excitable chat about the night ahead. I was a little nervous and tried to look cool (i.e. confident) – this became even harder when the two guys in front of us didn't get in. This intimidatingly beautiful girl with the bouncers at the front of the queue seemed to be choosing who was in and who was not, and she asked these two guys in front who they were with (meaning, I suppose, 'where are your partners?') and when they said that they were by themselves she just said 'Sorry boys, you're not going to be getting in tonight'. Horribly embarrassing – in front of all those people! After an hour of rainy queuing I know I would have been totally destroyed by this, although of course you can't say anything with six or seven massive bouncers next to you. We were waved straight through, and John even managed to crack a gag with the 'picker' girl. I was relieved although I felt a little sorry for the two in front who were wandering off home in the rain, dejected. But my thoughts quickly turned to the scenes inside. The crowd were already going berserk.[16]

Getting into a club – joining a clubbing crowd – can be a mere formality, yet it can be also be an anxiety-provoking experience, with some clubbers comparing club entry to passing an exam: 'one needs to study the look, prepare the body and stay cool under pressure' (Thornton, 1995: 114). Negotiating entry can certainly demand the display of a 'correct' style, the deployment of a very specific set of practices and a successfully employed 'personal front' – 'the complex of clothing, make-up, hairdo, and other surface decorations he [sic] carries about on his person' (Goffman, 1963: 25). However, it would be a distortion to pretend that a clubber keen to gain entry can access any club given careful observation of what is required – as I suggested above, 'the sharing of paraphernalia is a weak basis for membership or solidarity' (Warde, 1994b: 71). Not only does the development of this 'personal front' require practice, but the social identity of the clubber needs to qualify in certain other ways as well. Like most other social settings, clubs have 'entry qualifications' (Goffman, 1963: 9) that must be met, and these entry qualifications extend beyond wearing the 'correct' clothing and the 'correct' deployment of body techniques (Mauss, 1979). In addition to enforcing judgements about coolness, bouncers on 'the door' at clubs can also reinforce wider societal prejudices in respect of ethnicity, gender and 'good looks' – some clubbers find entry more problematic than others, and not everything is as always as 'PLUR' (peace, love, unity and respect)[17] as many clubs and clubbers would suggest. The following extract would appear to support Sarah Thornton's proposal that, while clubs may accommodate alternative cultures, they 'tend to duplicate structures of exclusion and stratification found elsewhere. "Black" men in partic-ular, find themselves barred, or, more usually, subject to maximum quotas. This ongoing fact should not be forgotten in the face of the utopian "everybody welcome" discourses in which dance clubs are intermittently enveloped' (ibid.: 25)[18].

SUN: I'm a hypocrite about this in that if I get in [to the club] I'm less fussed, but I'm less of a hypocrite about it than some people I know. I don't like clubs with dress policies [right] – there was a time when…okay I'll give you a little anecdote…[uh-huh]. I'd been to Temple[19] the previous week 'cos a friend of mine had gotten in on the guest list and it was y'know, a very nice club inside. Can't fault it in terms of its architecture and the sound system and what have you. But I noticed like, y'know, me getting fast-tracked in on the guest list queue and seeing all the poor punters waiting in the queue outside and like looking at the looks they got and remembering the times I'd been on that side of the fence and remembering the looks I'd given the guest list queue, and anyway, I sort of forgot about it, and I got in, and I noticed that the proportion of white people inside was like a lot higher than the proportion of white people in the queue. Ummm…and I noticed things like, even though I was bollocksed on an E, and everything was really happy and friendly, there was just something about it y'know? Guilty feet have no rhythm? Next week me and an Asian friend of mine, another bloke, we queued for ages, and we got to the front and what happened was they let us wait there at the front and then they opened the gate up as if to let us in and then out of the blue they said 'Right this way gentleman you're not getting in tonight'…now I know what the Temple bouncers are like so we didn't fucking complain, we just walked off totally quiet and gave them no shit at all. And as we walked off they called out 'You don't have a clue do you?' and I thought that was out of order. And me and Roger both smelled ourselves, it wasn't our clothes, y'know we were dressed right, it was 'cos we were two Asian men and they don't want too many Pakis in the club – they don't want ANY Pakis in the club. What we did was we jumped in a cab and we went to The Tunnel which has got a reputation for having a really strict door policy and we got there about 2 a.m., and there were a couple of beautiful people types trying to blag their way in saying 'Nah – we're on the guest list' and the bouncers weren't having anything of it and we were like waiting for them to like sort themselves out and eventually my mate called out y'know 'YO! look are you like going to let us in?' and the bouncer said 'are you going to pay?' and I said 'well, yeah, that's traditional!', and they let us in – bang! And that was cool. I think it's possible to have a dressed-up club without a dress code. I think you could just say to people can you make an effort to dress up and let them in first come first served. I don't like the elitism, I don't like guest lists, I don't like any of that shit – it's just an excuse for all the standard prejudices that you see in society to rear their ugly head in a scene I really like. So I try and avoid places with door policies.

In this extract, Sun is narrating the trials of gaining access to the same club (the Temple) to which I accompanied John in the extract from my research journal (p. ix). In fact, Sun and his friend, Roger, could quite easily have been the two who were excluded in front of us, except that they were 'white'. Sun's experiences are extraordinary in terms of the other clubbers that I interviewed, although others did mention instances of 'problems' at the door, in that Sun's identity as 'Asian' seemed to him to supersede his credentials of coolness. His 'belongings' were always about much more than just passing as 'cool'. Sun experiences the Temple in very different ways during separate visits – initially, he is fast-tracked in on the guest list, then, a second time, he and his friend, Roger, are point-blank refused entry. Second, Sun obviously blames his exclusion from the

Temple on his 'Asian' identity, proposing that door policies are often 'just an excuse for all the standard prejudices that you see in society to rear their ugly head in a scene [he] really like[s]'. How far this is actually the case is not clear from my clubbing visits and interviews, given the relatively narrow range of social identities displayed by the clubbers that I interviewed.

What is apparent from this extract, however, is that experiences of door policies and of gaining entry to clubs differ, even for the same clubber, and from club to club, and night to night. The identities of the clubbers attempting to gain entry to the club, and thus also to sensations of crowd belongings beyond the door, are strongly implicated in the processes of distinction that the door staff and 'pickers' exercise at the moment of entry. These door staff act as a 'last measure' in the processes of self-selection and self-identification that clubbers go through (Thornton, 1995: 22).

> If access to information about the club and taste in music fail to segregate the crowd, the bouncers will ensure the semi-private nature of these public spaces by refusing admission to 'those who don't belong'.
>
> (Thornton, 1995: 22–24)

The explicit use of supposedly 'inappropriate clothing' as an excuse to exclude someone from admittance to a club is common, whether or not that is the genuine reason. While, as Sun outlines and as I record in the extract from my research diary, exclusion for not displaying the correct identity can be based upon crude notions of ethnicity and gender, as well as upon the more obvious criteria of age. There are also instances where clothing really does appear to be a measure of distinction for the door staff.

BEN: Have you ever been told that you couldn't come in for any reason?

MARIA: Yeah, actually...about two months ago, yeah I was very upset [really?].
We went down to The Studio, and ummm...it was, I'm disgusted actually, I had nice shiny trousers on, I had my hair nice, with a bit of make up on, and I've got a pair of Vans [trainers], they were a bit dirty but they're 'fashion trainers', they're not 'trainers' and my friend was wearing a shitty old pair of jeans, a pair of fucked-up old Cats [shoes], a t-shirt with blim-burns in[20] and we got to the door and the blokes like looked at me and said 'Oh, I don't know if you can come in you've got trainers on'. I went 'oh, hello?' and he said 'well he can come in but I'm not sure that you can – I'll get the woman over' and she came over and she looked me up and down and she said 'no, you're too scruffy' and my friend was tidy enough and better presented because he had his Cats on and they weren't trainers, that he was allowed to go in, and then, while we were deciding whether we really wanted to go

back to get into some shoes so that we could get in here where we really didn't like the door policy, they let a little girl in, I say a little girl – I don't know how old she was, but do you know what I mean – a pathetic little girl with a little dress and she was wearing trainers right in front of me and I was like, well if you're going to have a door policy like that then stick to it. They didn't advertise anywhere that they had a no trainers policy, which I think was quite bad. You know, in the end I don't like that door policy, it's uhhh…basically the woman didn't like the look of me, I don't know why. It's like there were people behind us who were like office workers, out on the piss, not clubby types at all [right] but they were let in, and it's like what do you want for your club – do you want people that like music and enjoy themselves, or do you want people that are going to get smashed? I wouldn't ever go back there. It's a difficult one, door policies, because obviously if people don't like the look of you then obviously they're not going to let you in. I do find it quite difficult because I'm sort of…a lot of clubs do suffer from a male-only, sort of clubber type thing, and ummm…I'm not an unattractive woman and I'm very keen on having a good time and I think that most of the time I think I'm quite a useful asset to their club.

Rejection hurts, whatever the excuses given by the door staff. The negotiation of the door is a very explicit form of 'identity-exposure' in which clubbers attempt to pass a test. Many clubs have no dress codes, no restrictions in terms of the identities of those that may enter, with the exception of the legal age limit, which is difficult to enforce in any case, and getting into these clubs presents few problems other than finding the cost of admission. Other clubs, such as The Studio, to which Maria failed to gain entry and the Temple from which Sun was excluded, do discriminate in terms of dress, gender and even notions of how much a clubber might spend once in the club.

One of the clearest points that Maria makes in recounting her failure to get into The Studio is in distinguishing between 'office workers, out on the piss, not clubbey types at all […] people that are going to get smashed', and those like herself – 'people that like music and enjoy themselves' and yet cannot understand why those who run The Studio do not make a similar distinction. This distinction parallels the one made by Mark when he repeatedly refers to the 'tourist map' of clubbing in London, and how he does not want to go to the same clubs as these 'tourists'. Likewise, Maria infers that these are not the 'real' clubbers, and yet they are allowed in – quite possibly because they are likely to spend more money at the bar – and she is turned away. Maria is suddenly not very keen on The Studio when she sees who is being turned away and who is being allowed entry: 'they let a little girl in…'. Furthermore, this rejection leads Maria to look at herself, to wonder why she was not admitted and how she might not have fitted – there is a palpable feeling of distress in what Maria says towards the end of this extract.

CLUBBING CROWDS AND BELONGINGS

As a semi-private, musical environment which adapts to diverse fashions, proffers escape (sometimes with added transgressional thrills) and regulates who's in and who's out of the crowd, the dance club fulfils many youth cultural agendas.

(Thornton, 1995: 25)

These negotiations of the door, then, are indicative of the wider relationships and tensions between belongings and distinctions, identifications and exclusions, the cool and the 'mainstream' and inside(rs) and outside(rs) that have surfaced throughout this first stage of 'The night out'. The door becomes merely the last of many pre-clubbing tests of belonging, most of which are self-posed. From first going clubbing, through choosing a style, identifying with a certain crowd, defining a 'mainstream', selecting a night or nights, to queuing and passing through into the club, the clubber is engaged in reflexive processes of self-selection and self-identification. For some these processes apparently become unnecessary – they know where they like, they like how they see themselves and the crowd of which they form a part and they know they are guaranteed a 'safe night'. For others they provide moments of uncertainty and the risk of public embarrassment, as exemplified in Maria's story.

Yet the trials of the door appear more than worth paying. A clubber queuing to enter a club has already decided that they are 'right' for the club and think that they *could* belong. The door becomes the moment when certain facets of the identity that a clubber has constructed for himself or herself are used to gain access to a social situation in which other facets of that clubber's identity may be – relatively speaking – submerged or unshackled. Thus, these early and pre-clubbing stages of the night are significant because they provide a test of identity, and the passing of this test – the validation of a clubber's identity on the way through the door – can act to establish and reinforce the emergent belongings through which the clubbing crowd inside is bound together.

Notions and sensations of belonging – of the relationships between identities and identifications – are central to clubbing. The consuming crowds of clubbing are constituted, even in advance of their coming together, through concerns with shared styles, notions of the cool and the 'mainstream', and notions of belonging and exclusion. It is to the apparently chaotic yet intricately ordered interactions of the clubbing crowd once inside the club, through which these belongings are most powerful experienced, that I now turn.

SEB'S CLUBBING JOURNAL (06.05.95): It's strange that while every-body's happy to chill in Bello [a bar] and kill some time discussing last week, last night and how good tonight will be, really we're all desperate to get in the Q. It's struck me that the Party Posse are an extraordinary bunch indeed (this we already know – I've said it many times before). But what makes us special is our approach to life (speaking for myself, clubbing [...] is my life – and I'm sure the same is true for you). Cast your mind back to a time not so long ago, before all *this* started – when you'd go out for a night, I suspect that's all you did...not any more. Now we don't just go out for a good time, nor even the best of times – we go out to put on a show. Tonight, just like every Saturday night, the curtain was set to go up on another superb performance...

THE DANCER FROM THE DANCE

The musical and dancing crowds of clubbing

Every part of a man [sic] which can move gains a life of its own and acts as if independent, but the movements are all parallel, the limbs appear superimposed on one another, and thus density is added to their state of equivalence. Density and equality become one and the same. In the end, there appears to be a single creature dancing, a creature with fifty heads and a hundred legs and arms, all performing in exactly the same way and with the same purpose. When their excitement is at its height, these people really feel as one, and nothing but physical exhaustion can stop them.

(Canetti, 1973: 35–6)

INTRODUCTION

Clubbing is very much a social phenomenon at the heart of which is the clubbing crowd. The clubbing crowd is the foundation for the establishment of the belongings and identifications that I discussed in the previous section, as well as being the forum for the most overtly bodily enjoyment of the music and movement which largely constitute the clubbing experience. The clubbing crowd is thus central in any understanding of more complex and, in many senses, subtler processes of ecstatic pleasure and individual and group vitality (the subjects of the third and fourth sections respectively).

In this second stage of 'The night out' I build on the focus on belongings, outlined in the previous section of 'The night out', by developing a critical understanding of this clubbing crowd. Specifically, I am interested in the practices and spacings constituting the simultaneously musical and dancing crowd which lies at the heart of clubbing. Understandings of these practices and spacings – how, where and when clubbing is actually experienced in practice – can provide significant foundations for the unpacking of the clubbing experience in more explicitly emotional and imaginative terms later on in 'The night out'.

70

This section of 'The night out' is split into four sections. First, I briefly high-light the emotional and empathic quality of many crowds more generally. I propose that the experiencing of these crowds can provide pleasurable sensations of 'in-betweenness' – or *exstasis* – as crowd members flux between awareness and sensations of their own identities on the one hand and the identifications and belongings achievable through the crowd on the other. Second, I turn to the partially musically constituted nature of the clubbing crowd. I note the relative neglect of our musical experiences by the social sciences, the centrality of music to the practices and spacings of clubbing, and the notion that the listeners also partly create the experience (and production) of that music. Third, I pick up these points about music in directly addressing the dancing crowds of clubbing. Again, I note the reticence and/or inability of both clubbers and academics to discuss dancing. I then return to the sociality and performativity outlined in 'The begin-nings' in stressing the social and technical practices that constitute dancing. Fourth, I further elaborate this understanding of dancing in outlining what I refer to as the 'spacings of dancing'. I note how dancing is comprised of imaginative and emotional practices that are also simultaneously bodily and interactional in nature.

CROWDS AND TOGETHERNESS

> There is nothing man [sic] fears more than the touch of the unknown…it is only in a crowd that a man can become free of this fear of being touched…as soon as a man has surrendered himself to the crowd, he ceases to fear its touch.
>
> (Canetti, 1973: 15–16)

> [T]he age that we are about to enter will in truth be the ERA OF CROWDS.
>
> (Le Bon, 1930: 15; emphasis in original)

Individual clubber's relationships to the clubbing crowds of which they are a part, to the music and, perhaps most of all, to themselves, and the role of the micro-spacings of clubbing within these relationships, are of paramount importance to the clubber's enjoyment of the experience. The crowd-based nature of the clubbing experience appears critical in any understanding of its constitutive practices and spacings.

The notion of the crowd as in some way superseding the individual, as enabling the widespread experience of loss of self (*exstasis*) among those within and of it, is the point of departure of Canetti's (1973) discussion of crowds and power. For Canetti, crowds are about *belonging* and the power that this belonging can provide and instil. The boundaries around and between crowds – real and/or imagined – are crucial to feelings of belonging and identification. Canetti, for example, suggests that those 'standing outside do not really belong' (ibid.: 17), while for those inside sensations of belonging can be prolonged even beyond the physical experience of the crowd:

[The crowd] is protected from outside influences which could become hostile and dangerous and it sets its hopes on repetition. It is the expectation of reassembly which enables its members to accept its dispersal. The building is waiting for them; it exists for their sake and, so long as it is there, they will be able to meet in the same manner. The space is theirs, even during the ebb, and in its emptiness it reminds them of the flood.

(ibid.: 18)

The punctuated and at times almost ritual nature of first the integration then the fragmentation of crowds is itself partly constitutive of the bond through which members of that crowd may consolidate their attachments and identifications. Even while not physically within the crowd, members of a crowd may identify with certain spaces or sites central to the crowd (see Hetherington, 1996; 1997), specific times, memories, paraphernalia and even others of – but not at that moment within – that crowd. The power of crowds can thus resonate long after their dispersal.

Through opening out a complex and at times rambling typology of crowd forms, Canetti (1973) offers an understanding of crowds that he himself admits illuminates little of the individual feelings or practices of those within any particular crowd. Despite not dwelling upon the constitution of crowds on an individual scale, however, throughout his 'classification of crowds' Canetti weaves a common thread – the overall domination of all crowds by emotion. It is through the sharing of emotion – or, to put it differently, the evolution of a group ethos or 'we-rationale' (Goffman, 1963: 96) – that the disparate individuals forming a gathering can become bound into a crowd, a 'single being' in which a collective mind, albeit transitory, is tangible (Le Bon, 1930: 26). As Simmel proposes, this crowd ethos can exert powerful emotional effects upon the constitutive members of that crowd when, at certain times and in specific situations:

the individual feels himself [sic] pulled along by "the quivering ambience" of the mass as if by a force to which he is exterior, a force indifferent to his individual being and will, even though this mass is constituted exclusively of such individuals.

(Simmel, 1981: 116; cited in Maffesoli, 1993a: xiv)

As I alluded to earlier, the crowd and the experience of being within the crowd facilitates and also partially constitutes many forms of contemporary consuming. Indeed, Michel Maffesoli (1995) suggests that contemporary social life is increasingly marked by memberships of a multiplicity of social groupings, gatherings and other coalescences, all of which overlap and in some way contribute towards notions of our own and others' identities.

> In all these instances...a form of being-togetherness is lived out that is no longer oriented to the faraway, toward the realization of a perfect society in the future, but rather is engaged in managing the present, which one tries to make as hedonistic as possible.
>
> (Maffesoli, 1996b: xiii)

Empathetic feelings resulting from and concurrently also constituting these instances of being-together take on a new prominence as a collective sensibility temporarily – and only at certain times and given the 'right' contextual conditions – supersedes the atomised individual. This communal ethic, which foregrounds notions of proximity and the sharing of times, spaces and symbols (Maffesoli, 1995), may be, I propose, particularly prominent in instances of the experiential consuming which I introduced in 'The beginnings'. The group condensations may be based within work affiliations, hobby groups, campaign groups, sporting events, consumer lobbies, musical and performance crowds or youth cultures (Maffesoli, 1995); in short, within any social situations where the behavioural codes implicit in civil interaction may be transgressed or temporarily superseded with those based on an empathetic and more subtle sociality. I am interested here in a specific form of group condensation which acts as a context for consuming, but which is also a 'manifest togetherness' (Bauman, 1995b).

> The purpose of this togetherness is being together, and being together in large numbers...the higher-than-usual physical density gestates a similar density of sensual impressions: the overflow of sights and sounds, a higher-than-usual level of sensual stimulation, but more importantly yet a condensed, concentrated stimulation – reaching the elsewhere unreachable pitch thanks not only to the massive volume, but also to the monotonous homogeneity of stimuli: the same colour scarves wrapped around a thousand necks, the same jingle or ditty chanted, the same words shouted out rhythmically by thousands of breasts, the same twists and turns gone through by thousands of bodies...togetherness of this kind is mostly about the unloading of the burden of individuality.
>
> (Bauman, 1995b: 46–7)

Clubbing is one example of this form of collective experience in which 'common sense, the present and empathy' are dominant (Maffesoli, 1989: 11; emphasis in original). Furthermore, throughout 'The night out' I argue that the clubbing experience can be understood as a form of togetherness in which a central sensation is one of in-betweeness (or *extasis*) – this is the flux between identity and identification to which Bauman refers. On the one hand clubbing crowds 'anonymise' due to the sheer quantity of co-present clubbers and the sensuous overload that can make sight, recognition and communication problematic. Yet, at the same time, clubbing crowds can also 'individualise' – something that Bauman does not mention in opening out his notion of 'manifest togetherness'. Through, above all, dancing, clubbers can trace unique paths through the clubbing experience, distinguishing themselves as individuals. Yet these practices and spacings of dancing (see p. 90) are overwhelmingly crowd based. Through movement, proximity to and, at times, the touching of others, and (crucially) a positive identification with both the music and the other clubbers in the crowd, those within the clubbing and especially the dancing crowd can slip between consciousness of self and consciousness of being part of something much larger. The social constitution of the crowd, the clubber's understanding of that social constitution, and the mental and emotional approach of the clubber to that crowd are of central importance in catalysing this in-betweeness. Furthermore, it appears that individualising impulses are not always founded on positive and pleasurable notions of bodily expression.

DWAYNE: At Eutopia the people were much better than The Shed, so I didn't really have a problem with the people…in the first interview I mentioned that it's a couple of singular things or events that [yeah] make a night good or bad and at that particular Shed night, I came in and this bloke from my college – whom I consider, just from the way he looks and walks and acts, to be a COMPLETE arsehole – was there [I laugh] and I thought 'fucking hell, I cannot go to the same club that this bloke is at'…That, in the first place, changed my conception of the place completely, I mean I was with my friends and everything was going right and we had taken some drugs, but that just flipped my mental state from feeling well…to feeling really miserable, and then I looked at all the other people and I thought 'yeah, these are real West End Saturday clubbers' and they were really boring and…then I contemplated on the music and then I realised that it's such and such music that I really like and now it's mainstream so much, and look what kind of crowd that it draws…it was a self-amplifying process, and the music is…umm…the key, but people are very important, and I think ever since that night I haven't really been to a techno club, I think that was like the closing of the techno chapter for me in some ways, so I'm going to Abandon next Saturday…that was a very bad experience. It's very close to Banana Split in the arches where No Room For Squares is.

BEN: Right, so it's people and music…what about the certain aspects of the atmosphere, such as…I remember you saying a while back that ummm...music sometimes made it so difficult to talk...that you could only talk to the person next to you...that's a strange thing...?

DWAYNE: Yeah, that's the way I like to be with people, I generally...the more people that I have to communicate with, verbally, then the less comfortable I feel, so one-to-one conversations or one-to-two conversations are my ideal form of communication really [yeah]. That's what for me makes clubs much better than bars...even if you would go to a bar with one person, particularly if you did not know them too well, yeah, in a club it's totally okay to umm...to not talk for ten minutes and then do something else, but you can't just sit at a bar with someone and not do something [no, that's right – that's a good point], and so it makes communication, at least for me, much easier.

It starts to become clear in this dialogue how deeply Dwayne thinks about clubbing, what it means to him and how it works for him. First, the people in the clubbing crowd *matter* – clubbing is not simply about selecting a musical genre and a club that plays it and automatically having a good night. It seems possible that potentially good nights can be spoilt by, to use a 'British Rail-ism', the 'wrong type of crowd'. Second, Dwayne draws out direct links between the music and the crowd, particularly in terms of the mainstream/underground debate/tension. Third, a great deal of reflection is undertaken by Dwayne, both during the night and at a later stage, although my suspicion that, at least in this instance, this may have been partly affected by the fact that Dwayne knew he was to be interviewed a second time was confirmed by him in the second interview. Fourth, the crowd and the music combine to create the 'coolness' of a place. This coolness is critical to Dwayne's enjoyment; a crowd that is in some way 'cooler than him' seems essential for an enjoyable night in which Dwayne can lose himself in the crowd, can temporarily anonymise himself. Fifth, the density of the crowd paradoxically constitutes a more comfortable setting for one-to-one conversations. The notion that to talk in clubs is not always necessary – and is not always possible – is a feature that appeals to Dwayne. To talk becomes a choice rather than a social obligation. Finally, and perhaps most significantly, for Dwayne the 'wrong' type or style of crowd can lead to a *dis*-identification – a strongly individualising sensation that Dwayne is not – and is not able to feel – part of the crowd around him. This contrasts with the crowd practices of many of the other clubbers, particularly in respect of dancing (see p. 85).

I now examine in some detail two linked and major facets of clubbing crowds that are especially significant in the establishment of these sensations of in-betweeness and which also build upon and progress the notions of belongings and sociality that I introduced earlier. In particular, I am interested in unpacking notions of identification through concentrating upon their practical and spatial constitution. In doing this, I discuss clubbing crowds first as musical crowds and second as dancing crowds[21].

MUSICAL CROWDS

As long as we choose to consider sounds only through the
commotion they stir in our nerves, we will never have the
true principles of music and of its power over our hearts.

(J.-J. Rousseau, *Essai sur l'origine des langues*; cited in
Cranston, 1983: 289)

The sound of silence – the neglect of sonic worlds

Music evokes different reactions from different people in different contexts, not just in the sense of value judgements about whether a piece of music or a musical experience is 'good' or 'not', but also, and perhaps more interestingly and more subtly, in the very way that different people set about listening to and in a sense *understanding* music in varying contexts. Music has a quality inherent in its performance and consuming, and often mediated through technologies of (re)production, that enables its presence to transform and even create social spaces (Cohen, 1995; Frith, 1978; Valentine, 1995). For many people music is a very important element in their lives – it is literally the soundtrack to their everyday living. Music (in its countless forms) provides both aural backdrops and foci for many aspects of social interaction, particularly those based in leisure, pleasure, entertainment and play. Yet, music as an object of intellectual enquiry has been neglected, with academics being peculiarly silent on music[22]. Our sonic landscapes remain overwhelmingly unknown, let alone mapped, with what Smith (1994: 233) calls the 'ideology of the visual' being afforded an epistemological privilege in social research[23]. Smith adds, 'the majority of human geography remains devoted to seeing the world, or speaking about it, rather than to listening or hearing' (ibid.). This neglect has been due to the comparative difficulty of the translation, notation and representation of music, musical experiences and musical understandings. While drawing, painting and even photography and film are relatively easily transferred, copied, translated and discussed, music exists only in the continual present of its unfolding. Thus, while the 'art' of music is in its making and the moment of its experiencing – whether producing or consuming, or both – the 'language' of music is usually found only in formal notation and the written musical text. A further and associated reason for the neglect of musical forms in intellectual enquiry has resulted from the attitude that music is not worth studying because it is less explicitly representational than literature, painting and the other visual arts. Music is not, or can appear not to be, proposi-tional, but just *is* (Storr, 1992). In one sense, music 'disappears' in the moment of its manifestation – it does not linger, but exists only in its being (Storr, 1992).

While peculiar intellectually, given that music is clearly so important to so many people, this silence on sound is thus not surprising in practical terms and is reflective of wider societal attitudes. The rationalist impulse that has dominated science and art in Western culture, arguably until relatively recently, has resulted in a failure to recognise fully the significance of the emotional within social and cultural life, especially where the emotional has an instinctive, ephemeral and at times elusive quality to it, as in reactions to music. The idealised and hypothetical notion of a rational thinking, yet somehow not feeling, subject that has prevailed has resulted in rationality taking precedence over the emotions, idealism over materialism, culture over nature and objectivity over subjectivity, with the mind and body often situated in binary opposition (Thomas, 1995). Whereas the spoken word has evolved into the dominant form of rational communicative action, music (originally an accompaniment to spoken communication) and its embodiment through dancing have come to be associated with the unconscious, the symbolic, the fleeting and the emotional, and thus largely ignored (see Storr, 1992). Susan Smith (1994) notes how the value of music is not recognised, even in Raymond William's (1982) study of culture and society, yet she adds that the benefits of an awareness of music for sound research lie precisely in the fact that it lacks 'the concreteness that so appeals to empiricism, [and] is more allied than vision to those emotional or intuitive qualities on which the interpretative project rests' (Flinn, 1992; cited in Smith, 1994: 232).

Music matters

The level and intensity of meanings invested in music by young people is unmatched by any other organised activity in society, including religion (Ross, 1994). Music exhibits strong relationships to social structures, 'not only repre-senting social relations [but also] simultaneously enacting them' (van Leeuwen, 1988). Music plays a central role in the constitution of identities and communities (Frith, 1996) with especially young people using music to situate themselves historically, culturally and politically (Frith, 1992), and, I would add, stylistically.

Simon Frith suggests that music 'provides us with an intensely subjective sense of being sociable…it both articulates and offers the immediate *experience* of collective identity' (1996: 273; emphasis in original). Music regularly soundtracks our search for ourselves and for spaces in which we can feel at home, through, for example, the use of Walkman personal stereos in 'self-produc[ing] [our] own space' (Valentine, 1995: 481)[24]. Yet, music also assists us in 'getting out of ourselves', in experiencing *exstasis* and momentary loss of self through musically supported social interactions such as clubbing. Music can assist in our construction of identities and identifications through the experiences it offers us of our own and others' bodies, of times and spaces, of sounds and images, of sociability – all experiences through which we are able to 'place ourselves in imaginative cultural narratives' (Frith, 1996: 275). On the other hand, music can also have the effect of intensifying shared experiences through magnifying an emotion or set of emotions that an event or social interaction brings forth by simultaneously evoking similar emotional and physical responses amongst a group of co-present people (Storr, 1992). Musical memories can be some of the strongest, most personal and most easily evoked. For example, Sara Cohen (1995) explores the relationship between music, memories and places in her study of the biography of just one Liverpuddlian, 88 year old Jack Levy. Arguing that music not only reflects, but also creates places, Cohen proposes that in playing a 'unique and often hidden role in the social and cultural production of place', music foregrounds the senses through 'action and motion' (ibid.: 445).

> When Laura was here I had the records arranged alphabetically; before that I had them filed in chronological order, beginning with Robert Johnson, and ending with, I don't know, Wham!, or somebody African, or whatever else I was listening to when Laura and I met. Tonight, though, I fancy something different, so I try to remember the order I bought them in: that way I hope to write my own autobiography, without having to do anything like pick up a pen. I pull the records off the shelves, put them in piles all over the sitting room floor, look for *Revolver*, and go on from there, and when I've finished I'm flushed with a sense of self, because this, after all, is who I am. I like being able to see how I got from Deep Purple to Howling Wolf in twenty-five moves; I am no longer pained by the memory of listening to *Sexual Healing* all the way through a period of enforced celibacy, or embarrassed by the reminder of forming a rock club at school, so that I and my fellow fifth-formers could get together and talk about Ziggy Stardust and *Tommy*.
>
> (Nick Hornby, 1995: 52)

Music also assists in the constitution of bounded places through, for example, national anthems by 'using sound to communicate what cannot be spoken' (Smith, 1997: 517) and solidifying senses of the self and of others (see Kong, 1995a; 1995b). Music can thus be a way of telling stories about oneself, often to oneself. Music also assists in more dynamic and explicitly process-constituted senses of spacings and histories. The presence or absence of music can totally transform spaces, and the memories and imageries of music can mark out a past history in extraordinarily powerful ways, for example in Nick Hornby's notion of his changing and developing self over time.

Furthermore, the social nature of music – the emphatically empathic impulse that it can instil – led Nietzsche to propose that, of the arts, music was one that 'so sharpened our sense of participation within life that it gave meaning to life and made it worth living' (Storr, 1992: 150). Participation in producing, reproducing and consuming music can act to re-state our sense of vitality, our thirst and enthusiasm for living, despite other aspects outside of the musical experience being sources of anxiety or uncertainty. One flip side of the commercialised and commodified global culture of music, of which Adorno (1973; 1976) is perhaps the best known and most vehement critic, is the potential that music offers for a localised and popular sense of resistance through the numerous contexts of both its production and consumption[25].

The work of Will Straw (1991; 1993) on local rock scenes in the USA, Ruth Finnegan's (1989) study of amateur music-making in Milton Keynes and Halfacree and Kitchin's (1996) retrospective on the Manchester music scene of the early 1990s all intersect in their recognition of the very *local* sense of musical production that is in operation. They also recognise that this sense of locality and of being 'unique' can enable predominantly white, youthful and insular audiences to assert their differences from a hegemonic mainstream[26]. This expressly musical experience of locality and the key role that music plays in the social and cultural lives of the young people concerned are indicative of the important ways in which young people's identities and identifications are linked to the modes, practices and spacings of their leisure activities. How, then, do the 'musics' of clubbing work within the context of the clubbing crowd?

The social centrality of music to the practices of clubbing

The clubbing experience is primarily about music and the clubbers' understandings of that music. These understandings are largely expressed and addressed through dancing, to which I turn in detail on p. 85. In 'The beginnings' I noted how clubbing is an overwhelmingly urban and predominately youthful experience. However, what characterises clubbing above even these qualities is this *musical* nature. Music is central to the clubbers' enjoyment of their clubbing experiences (a point which presents itself forcefully in the accounts of clubbing that I have collated). Furthermore, through the identificatory power that very loud music exerts over the clubbing crowd, an emotional bond (or ethos) can be generated and reinforced between what is effectively a large group of strangers. This power of music to unite ostensibly disparate individuals, particularly when combined with other emotive environmental features or events, such as lyrical messages, can be as Storr (1992: 46) proposes, 'terrifyingly impressive'.

In one sense music can be presented as the essence of clubbing, to be what clubbing is about and what clubbers unite around or through. Revisiting and revising Shields' (1992a) and Hetherington's (1996) notion of the social centrality of place, it could be suggested that in the clubbing experience, rather than a fixed notion of a place or a site being focal, clubbers come together around the 'musics' of clubbing. These 'musics' act as a focus for the articulation of identities, the development of a sense of belonging and ultimately facilitate an identification for many of those within the clubbing crowd. Clubbers often define themselves in terms of their preferred music(s) and the associated crowds thereof – 'Do I want to be like them?'; 'Do I want to be associated with that crowd?' – and the stories of Dwayne and Mark (pp. 55 and 58) subtly and differently illustrate this point. Blacking (1973) and Spencer (1985) also note this totemic aspect of music, particularly when it is combined with dancing. Music can be seen – or rather, heard and experienced as:

[a] visible (in this case danceable)...symbol of the invisible force of the collectivity...active within the group as a whole, an expression of senti-ments that well up in the individual in response to the established collective symbols of his [sic] society.

(Spencer, 1985: 15)

SEB: It's not the drugs, it's not the party scene, it's the music and like I've said to you before, the club is just an environment that allows you to listen to it the way that it should be listened to – LOUD music. Like-minded people and loud music equals a good atmosphere, but I buzz off the music in the car, wherever, and it influences me so much, like I said, in the clothes that I buy, in the way that I think and I think that taking drugs and going clubbing has just coincided with me adopting a...I mean I've always been enthusiastic and passionate about things that I enjoy and I've never enjoyed anything more than this, so I'm a very positive person and I think taking drugs has opened my eyes to another kind of scene, a less inhibited scene, a scene away from the aggression that is associated with alcohol and other such clubs, you know, what you get with alcohol. I think, again because of my personality and the way that I am, a very emotional person, a very tactile person, I think it suits me down to the ground [um-huh]...

Three main points can be taken from this short dialogue. First, for Seb the music comes before all other aspects of the night, and in fact transcends the night to impact on other contexts of his everyday life, for example his 'travelling times'.

Second, from an *initial* identification with the music and the crowds that are associated with the particular genre of clubbing music he likes, Seb has branched out
so that the music impacts upon his style and appearance, his way of thinking more
generally and especially other important aspects of his life as a person, such as
drug consumption. Third, the complex interconnectedness of the relationships
between music, clubbing, drug use and lifestyle is evident. Within a few sentences
Seb touches on all these aspects of his life without interruptions or promptings,
seamlessly snowballing from one to the next.

The blurring of the distinction between music and its context or environment
and the (con)fusion of clubbing with other aspects of social life, both of which,
for example, Seb talks of on p. 80, are characterised by a concomitant smudging
of the distinction between the visual and the aural. Lights, lasers, mirrors, smoke,
darkness, motion – these can all be important elements in clubbing, resulting in
what one *sees* as always also being an effect of what one *hears*:

> [S]pace becomes movement as dancehall, club, and warehouse are
> shaped by the dancing bodies that fill them; when silence falls, the
> setting disappears. The dancers are performers, programmed by the
> deejay [sic]; the music stops, play time – the *scene* – is over.
>
> (Frith, 1996: 156; emphasis in original)

BEN: What about music?

SUN: I was into the music first, and I'll be into the music always. What drives it is
the social infrastructure…music is like the…focus, it's what you talk about.
You go to a club and the first thing you ask is 'What was the music like?'
[yeah]. If you look at uk-dance [an Internet mailing list], there's very little
about dance…it's usually about music 'cos that's what the clubbers, that's like
the central point of reference, y'know, that people had. And it's also something that you can get into quite easily because you can buy the magazines
and you can buy the records and you can have an opinion on something…it's
easily shared. E's are difficult to score and cost a lot of money, but
music…I've always listened to music, I used to listen to Janice Long and John
Peel, used to buy *Melody Maker* religiously, y'know, switched to dance,
converted to dance in '88, and I like all sorts of music…

This story evokes the centrality of music for Sun, particularly given the difficulty of talking about dancing – a point that I unpack on p. 89. Further, the
accessibility of music for a whole range of purposes is evident: as something that
can be shared, as the foundation for social interaction, as a topic in which expertise can be gained and demonstrated in the construction of identity and
identifications, and out of which a sense of belonging can be forged.

Making music together

In his seminal anthropological text, *How Musical Is Man?* (1973) – one of the first in modern Western musicology/anthropology to explore some of the relationships between 'musics', cultures and societies – John Blacking emphasizes the musical abilities that we *all* have, proposing that listeners are in one sense no less talented as 'musicians' than those who have produced the music to which they are listening. The labelling of the majority of the population as 'non-musical' while at the same time a tiny minority are afforded the status of 'musicians', is, Blacking suggests, a gross misrepresentation based upon flawed concepts of supposed expertise and a conflation of two quite separate points.

First, the techniques of musical production should not be confused with the techniques of proficient listening. An individual who has not been formally trained in music can display expertise in the form of recognition and knowledge of a piece of music, and further – and perhaps more significantly – that individual may well enjoy the music no less than another individual who has been formally trained. Second, performed music means nothing without a musically adept audience; that is, without an audience who respond to, and distinguish between, different sounds and sequences of music, the performance of that music would be pointless. To take, as Blacking does, the example of a classical performance:

> [T]he very existence of a professional performer, as well as his [sic] necessary financial support, depends on listeners who in one important respect must be no less musically proficient than he is. They must be able to distinguish and interrelate different patterns of sound.
>
> (Blacking, 1973: 9)

An excellent example of Blacking's argument – that the audience engage in the practices of what he calls 'creative listening' and that the performance depends to a greater or lesser extent upon this involvement – can be found in the practices of clubbing and of the engagement of clubbers with clubbing music through dancing. The sense that the 'audience' at a club night are actively engaged in the *production* of that night, as well as in its consuming, is another of the essential qualities of clubbing. Dancing, in these terms, is a prominent form of creative listening, relying on shared knowledges of music, common appreciation of certain musical forms, and distinctions between the many differing forms of music on offer. It has been argued that without the 'audience' of clubbing – that is, the other members of the clubbing crowd – clubbing would not exist; clubs without clubbers lose their meaning (Brown, 1997).

In clubbing, while the music is usually produced elsewhere, the DJs *re*produce it through the use of technology. Yet, the music is also partly reproduced through the clubbers and their role as the 'audience', as an active and 'musical' crowd, listening to, understanding and expressing the music through themselves and their dancing. The clubbers as consumers of the music are also simultaneously the producers of the performance of that music. There is a collusion between the clubbers as audience and the clubbers as performers, a kind of 'knowingness' between members of the crowd and their understandings of the music (Frith, 1995).

Listening to and understanding clubbing music

Acting a part,
Living a film,
Alongside the scene
plays an inner melody...
The Soundtrack of Life.

(The Gentle People, 1997)

Many clubbers describe a sensation experienced while clubbing of being in a state or in a place that is in some way removed from the 'normal' times, spaces and social relations of their everyday lives, of being in a realm of fantasy, fun and freedom[27]. These sensations can be traced back to the ways in which the clubbers listen to and understand the music in the clubbing experience. What we feel about music is essentially what the music *means* to us, and so the clubbers emotional and physical reactions to music are integral to uncovering the meanings which they invest in music. In the context of the crowd interaction, which characterises clubbing, the emotional and physical reactions of the clubbers to the music can coincide in the evolution of an ethos or identification. This can occur even in the presence of differences that might normally preclude this sharing of emotional space; for example, differences in age, ethnicity, social class or gender[28].

Intellectually, the emotional impacts of listening and responding to music have not been widely discussed, with more formalised and mechanical listening practices being given priority. Simon Frith (1996) suggests that there are different modes of listening – what he calls 'ideologies of listening' (ibid.: 142) – and the mode of listening that 'demands' this distanciated, almost purely aural understanding of music is a product of the rationalist approach to music and to art more generally. Arguing that any description of the performance of a piece of music should be strictly formal and neither remotely evocative nor emotional, Igor Stravinsky (1962) proposed that straying into the sphere of the emotions would be to contaminate the music in its 'original' and 'pure' form. Frith (1996) notes how this approach to listening (based upon practices of 'structural cognition') typifies the high cultural ideal of the last 200 years or so and is far removed from the everyday listening practices that most of us utilise in our enjoyment of music, and is thus useless in terms of developing any understanding of these everyday practices.

> In musical terms, which is the odder event: a classical music concert where we expect to see musicians bodily producing the music which we listen to thoughtfully, silent and still; or a club night at which we don't expect to see the musicians (or even the deejays) producing the sounds, but in which *our* physical movement is a necessary part of what it means to listen?
>
> (Frith, 1996: 142; emphasis in original)

Whilst the point he wishes to make is clear, Frith's rather neat dichotomy between two quite different modes of listening overlooks the significance of events and situations where the boundaries between these modes are blurred. For example, the annual Proms Concerts at the Royal Albert Hall where 'revellers' (never merely 'listeners' at The Proms – a notion that is interesting in itself) traditionally not only stand and dance (if somewhat minimally), but even 'dare' to cheer, shout and sing – they actively embody the performance in many ways as much as the orchestra on the circular stage.

Strong emotional reactions can be provoked without words or lyrics, either within the music or between co-present clubbers, although this by no means infers an absence of language on, and off, the dance floor. The techniques of bodily communication are a language in themselves, a feature of the dance floor that challenges Hemment's (1997) assertion that the 'primacy of vision' has been displaced by that of the aural in clubbing. A more accurate assertion would be that the primacy of vision is submerged at certain times and in certain places and for certain people more than others.

Thus, in contrast to Stravinsky, I propose the foregrounding of a mode of listening that prioritises the simultaneously *motional* and *emotional* understandings of the listeners. Listening as an embodied and emotional activity takes on varying forms in various spaces. Some spaces are more suitable for a more fully embodied understanding through dancing than other spaces, and I return to these spaces on p. 90 where I discuss dancing. Some sites of musical performance and entertainment actively promote a particular form of listening, whilst others discourage or even forbid 'unsuitable' bodily responses to the music. Still others blur the boundaries between forms of listening context. The spaces of musical production and reproduction have always been contingent on the particular form of music and the mode of listening employed within those spaces (Frith, 1992; 1996). Musics that demands a relatively static and emotionally restrained audience have always occupied specific places – galleries, country houses, concert halls, churches – while at the same time popular musics (what Frith terms 'low musics') have conventionally been associated with bodily expression, pleasures, carnality and emotions, and thus linked with sites of deviance and depravity: the jazz bar, the brothel, the gig...even the city itself. This has certainly been the case with the musics of clubbing (Frith, 1996; Hughes, 1994; Ward, 1993; Welsh, 1993; 1996).

BEN: Why house music then?

BRUCE: Someone once said that, what was it? Talking about music is like dancing about architecture...incredibly difficult! It's bloody hard, umm...I like music and umm...I've got to the stage whereby I don't really listen to music to get depressed whereas when I think you're a teenager you sit in your room a lot with the Cowboy Junkies or whatever playing and you want to slit your wrists and I don't want that – I like the way that it can make you feel elated, in which it can lift you, and so I like house music and the way in which it combines that with lyrics and just things that are just constantly lifting and interesting. I just find the way house music is put together to be interesting and to appeal to me and to continually appeal to me as well. I like the way that it combines disco and funk and I like all that, and I don't know, I just like dancing, I absolutely love it...

There are so many interesting points that spin out of this dialogue. Bruce clearly thinks seriously about his own relationships with the music and is able to evoke relatively concisely what music means to him. Bruce makes a clear distinction between how he *used* to listen to music as a non-clubber – sitting alone in a room, and how he listens to music *now* as a clubber – as interesting, as elating, as uplifting. Third, and connecting directly with the imminent discussion of dancing, Bruce discusses his listening practices through reference to dancing. For most clubbers the role of their bodies in their understandings of music is very important, as they seek to both individualise themselves and yet attain some sense of unity with others through their dancing.

DANCING CROWDS

Dance will bring the dead world to life and make it human.

(Marcuse, 1970: 132)

Dancing is the major expressive form through which music is understood and enjoyed during the clubbing experience. Together with and inseparable from the music, dancing is what clubbing is about and is largely what clubbers 'do' while clubbing. Through dancing to music in a crowd, a high degree of individuality may be generated, while concurrently providing, if the dancer so desires, and is technically competent, opportunities for the loss of that individuality within the crowd.

But what *is* dancing? Why has its obvious social centrality for many people been glossed over and what are the socio-spatio-temporalities of dancing in the clubbing experience? How is dancing performed and how might we start to conceptualise its practices?

Dancing in the dark

Dancing, where the explicit and implicit zones of socialised pleasures and individual desires entwine in the momentary rediscovery of the 'reason of the body'…is undoubtedly one of the main avenues along which pop's 'sense' travels.

(Chambers, 1985: 17)

In Brazil, there are certain dances which, in effect, write their own meanings. And there is an understanding of the possibility of a corporeal intelligence. There are things I learned in Brazil with my body, and some of these things it has taken me years to learn to articulate in writing. But that is not to say that they were without meaning when I could only speak them through dance.

(Browning, 1995: xi)

Dancing is a mode of behaviour in which the relationship between movement and thought (or motion and emotion) is central, even if this relationship can be difficult to evoke for the dancer and may be problematic to define or explain for the researcher (Hanna, 1987). As well as being a social, political, psychological, occasionally pharmaceutical and often economic activity, dancing is primarily a form of communication or body language and a mode of expression, a performance in which verbal communication can be supplemented and even temporarily superseded, either intentionally or through compulsion. I agree with Boas (1972; cited in Snyder, 1974) in proposing that ordinary gestures and actions can become 'a dance' if a transformation takes place within the person – a transformation that takes that person out of his or her 'ordinary world' and places them instead in a world of heightened sensitivity and altered perception of self, others and/or the environment.

I thus see dancing as a conceptual language with intrinsic and extrinsic meanings, premised upon physical movement, and with interrelated rules and notions of technique and competency guiding performances across and within different social situations. Dancing within clubbing might be conceptualised as an expressive form of thinking, sensing, feeling and processing which may be constituted through, as well as reflecting, strong relationships between a clubber and the clubbing crowd, and in turn between that clubbing crowd and the society of which it is a part. Dancing can be a form of sexualised ritual (McRobbie, 1991), a form of expression (Storr, 1992), a kind of exercise (McRobbie, 1994b), a form of individuation yet also one of unity (Frith, 1996), a language (Shepherd, 1991), yet a language that is non-textual (Frith, 1995). Dancing within clubbing can be about fun, pleasure and escape, about being together or being apart, about sexual interaction or display, about listening to the music, and even a form of embodied resistance and source of personal and social vitality (as I discuss later in 'The night out').

Traditionally, dancing of whatever form – ballet, modern, folk, popular – has been neglected by the social sciences which have tended to approach dancing as primitive, carnal and thus unworthy of study (Frith, 1996)[29]. Routinely presented as being based purely in bodily pleasures and thus as somehow depraved, dancing has also been theorised as a form of 'ritual possession' (Ramsey, 1997), as signifying an unbalanced mind (Backman, 1952; Hecker, 1970) and as in opposition to the practices of more formalised musical appreciation. Judith Lynne Hanna explains this long-standing avoidance of any critical engagement with dancing by highlighting a combination of 'Puritan ethics, social stratification, concepts of masculinity and a sense of detachment from nonverbal behavior [sic]' (1987: 9). The apparent playfulness of dancing placed it as diametrically opposed to a Puritan ethic buttressed upon the centrality of work and grafting. Thus, dancing was considered unsuitable for attention, not even in the guise of a form of deviance.

As a playful activity, then, popular dancing in particular is seen as lacking in 'serious' content. Nigel Thrift paraphrases Ward (1993) in pointing out how what he calls this 'peculiar invisibility of dance' is partly due to the fact that the practices of dancing lie well beyond 'the rational auspices of Western societies' (1997: 145). Interaction based upon and within verbal interaction has historically been in the ascendance in the social sciences, with sociologies of sports and other forms of leisure, such as dancing, seen to demand little in the way of rigour, instead being 'the hobby of anti-intellectuals' (Hanna, 1987: 10). If it holds the potential for fun then it cannot sustain or be worthy of serious study[30].

The reasoning behind the traditional neglect of dancing goes further than its associations with play and the body – both areas of social life with which anthropologists, sociologists and especially geographers have been slow to engage. Like music, dancing is easier to demonstrate than to talk about. Dancing is difficult to define and it is inherently a non-verbal language. Thus, dancing can be tricky to describe or even evoke (Blacking, 1975). There are also problems of notation and disputes over history, aesthetics and criticism (Brinson, 1985; Ward, 1993). More generally, when dancing *is* considered there is usually an emphasis upon stage 'dance'.

While the intellectual barriers to developing understandings of dancing lie firmly within the bounds of academe, the perverse attitudes towards bodily pleasure of which these 'anti-dancing' prejudices are just one facet are, again, firmly rooted within society. Despite the close association between music and dancing, it is telling that many more parents want their children to study music rather than dance (McRobbie, 1991), a bias further demonstrating the connotation of dancing with the carnality of the body (Frith, 1996; Storr, 1992). In addition to this there have also traditionally been concerns about the female display that occurs as girls dance, a sense of alarm that is rooted in a more general ambivalence over sexual display and which is largely to blame for the fact that, for example, the 'early middle-class pioneers of the youth club movement discouraged dancing as a legitimate club activity' (McRobbie, 1991: 194). Further, the activities of dancing and the spaces associated with them have provided ample ammunition for moral panics about young people in society, although McRobbie notes how it is usually less the dancing itself that is to blame than the fear of associated activities, such as drinking, drug use, underage sex and the possibility of violence.

SUN: I think that explains the male bias on uk-d [Internet mailing list], 'cos women talk a lot more about the dance scene than men, y'know [umm-huh]...it's just a very obvious sociological observation that you don't get as many women trainspotters[31].

BEN: So what...if people enjoy it so much why don't they talk about it?

SUN: It's even harder to write about than music. I've been toying with the idea that next time someone posts up a mail saying they can't dance to jungle posting up a sort of piss take, y'know, Sun's *Beginner's Guide* to dancing to jungle, y'know. 'Cos I've thought about this, 'cos I've got no fucking sense of natural rhythm – the reason I can dance is that I've been going clubbing for about eight years and I...and people are very self-conscious about it – I am very self-conscious about it. I have been told that I can dance reasonably well. It is just practice, ummm...and then...uh...

BEN: So you dance for yourself, but for others too...?

SUN: Oh yeah, you want to look horny, and I go off dancing on my own quite happily, and then part of it is just this display thing, part of it is just like uh, a sort of martial arts sort of thing of control [ummm, yes] and the other thing is you do get an amazing rush. Basically, things really took off once I learnt how to move my feet, um, and I've always said that a lot of people that are into techno, and diss garage for being slow, they don't actually know how to dance – they don't see why people like garage [no] ummm...

BEN: Okay...what do you think of when you dance...if anything?

SUN: Oh, I listen to the music, I look around, I watch how other people are dancing, sometimes I just close my eyes and get basically to almost meditative, I try to cut out everything, I concentrate on how my feet move, I think a lot about my hands.

The practices of dancing and the practising of dancing do matter, even if clubbers, like academics, also have difficulty articulating their dancing experiences. For Sun, there are clearly tensions between expression and control and between display and meditation. In reply to the question that I posted on the Internet mailing list, uk-dance, in February 1996, 'why don't people talk about dancing on uk-dance?', I received dozens of replies. Most of them were along the broad lines of this one:

> Dancing is impossible to discuss without sounding wanky. So is music for that matter, which is why we end up talking about drugs the whole time...really though, dancing to dance music is more-or-less un-choreographed – it's a gut-level reaction/interaction with the sound (see what I mean about wanky?). That's why choreographed dancing, as on TOTP [Top of the Pops], is so contrived + silly looking. There is no structure to attach words to – so we can't talk about it.
>
> (contributor to uk-dance, February 1996)

The 'meanings' of dancing are multiple and related to the larger cultures and social structurings of which the dancing individuals are constitutive. Even to a single individual dancing can assume varying forms and meanings at different times and spaces throughout and between nights out. Moreover, dancing can reflect powerful social forces that demand some explanation (Spencer, 1985), for the issues that dancing raises impinge upon highly significant issues, such as political control and organisation (Browning, 1995), the relations between 'race' and gender (Thomas, 1996), social psychology, social morality and resistance and domination (Brinson, 1985; Thrift, 1996; 1997).

In addition to these reasons, I propose that the practices of dancing as constituted through clubbing demand a deeper understanding for at least three reasons: first, because they form a foundation of what for many young people is an important form of social interaction; second, and building on this first point, dancing is based largely upon notions of spatio-temporally bounded belongings and group dynamics and is strongly implicated in processes of identity creation and in the formation, consolidation and fragmentation of identifications through the embodied and performed nature of its techniques and skills; and third because of this social centrality and these links to the formation of identities and identifications, dancing resonates powerfully in the dancers' everyday lives and thus in society: 'in a very important sense, society creates the dance, and it is towards society that we must turn to understand it' (Spencer, 1985: 38).

> Dance is…an active creation of meanings, that is, social action dependent upon social relationships at the time. This is the reality of dance whether it takes place on the village green, in a disco, or in an opera house.
>
> (Brinson, 1985: 211)

As we shuffle, then, perhaps somewhat belatedly, onto the rapidly filling dance floor, I now turn explicitly to these practices of dancing. Dancing is a complex activity and its intensity can overwhelm theorists as well as dancers. Approaching dancing through a performative lens is only one form of understanding. However, the performative lens does illuminate and place at centre stage the conceptual spacings of dancing – spacings through which clubbing more widely is also practically constituted and through which we might begin to understand better the complex nature of clubbing.

SPACINGS OF DANCING

> There is no social order, only modes of ordering.
>
> (Hetherington, 1997: 280)

> What is certain is that all of this relates to space.
>
> (Maffesoli, 1988a: 150)

> The organization of space is assumed to have an important bearing on human interaction and its dance manifestations.
>
> (Hanna, 1987: 200)

In the first stage of 'The night out' I mentioned how, within the clubbing experience, notions of the in-crowd and out-crowd, notions of coolness and mainstream, and thus of identifications (belongings) and dis-identifications (outsiders), were critical to the development of an understanding of those cultures, and yet always relative. In building upon this and also developing my earlier points about sociality and performativity, I now want look in detail at the practices, timings and spacings of dancing – the major constitutive practice of clubbing.

Dancing is a response to the music, to the clubbing crowd, and to wider aspects of the clubbers' social lives and identities. As Frith (1996: 139) suggests, 'the ideal way of listening to music is to dance to it, if only in one's head', and through dancing one may lose oneself in the music physically and mentally. One can dance to forget certain aspects of one's identity, to express happiness at being together, and to find relaxation (Hanna, 1987). In the clubbing experience dancing can be about losing control over one's body and yet somehow gaining a deeper level of control over the body through willingly yielding oneself to this process of relinquishment. Dancing can be about becoming part of and submitting to the dancing crowd, yet also individualising the self through the bodily practices of dancing within that crowd. Dancing can be about expressing oneself to others and constructing one's own notion of self concurrently. Dancing fuses notions of 'inside' (emotions) and 'outside' (motions) as the internal becomes externalised, and the external becomes internalised (Butler, 1990a).

BEN: When you were dancing then, did you think...?

VALERIE: I'll tell you what – dancing that night was quite amazing...yeah? I haven't had an E^{32} like that. It was really really dancey. I wasn't thinking about it – it was just happening to me [um-huh]. I was FORCED to do it. What I was doing I was being forced to do by something, yeah...

[short interruption by a sniffing dog – we're in a park]

BEN: ...tell me about this feeling of being forced...did you *want* to dance?

VALERIE: Oh yeah, I did want to do it, it was fun, it was just so...the fact that I was doing it...d'you know, the last time that I said to you, ummm...that you feel like you're the conductor – you ARE the music – you ARE the conductor...yeah? [yeah] – I had that yeah? I was just creating the music yeah, and that gives you a buzz. It just felt good. It just felt physically excellent.

BEN: Ummm...okay, so you're dancing and it's really hot – what about the others? You're kind of dancing in a little group...

VALERIE: Mike said that we should start dancing in a circle because then we get more room, because usually you're like facing the front like a fucking tribe yeah [yeah, that's right]. It's like such a personal thing though – you're not dancing for anyone else, you're dancing for yourself, you're dancing because you HAVE to dance!

I want to draw out three main points from this exchange. First, it is significant that in response to a question about *thinking* Valerie replies about the *physicality* of dancing. The tension between being apparently 'forced' to dance and feeling 'excellent' is discernible even in the short extract reproduced here, and it is a tension that is similarly evoked in what Valerie goes on to say about the duality of dancing for oneself, yet also in and for the group. Second, dancing seems to be about having fun, being in control and being with – and literally close to – others, yet at the same time, relinquishing that control, being compelled to dance and losing touch with those around you – closing up. Third, a suggestion of the important role of spacings and orientations in Valerie's inhabitation of the dance floor is evident in the last paragraph of the extract. It is wrong to assume that clubbers merely drift around on the dance floor. Their location, expression and bodily practices are all carefully monitored, often pre-arranged and usually practised. With further practice these spacings might become 'second nature'.

Pitched in terms of the relationships between performativity and sociality, dancing clubbers constantly both produce and consume the activity in which they partake. Dancing clubbers are continually investing and deploying skills, techniques and competencies (in other words, sociality) that they have acquired and are acquiring in the processes of attaining and maintaining notions of group identifications and social identities and in their own imaginative constructions of self. For dancing clubbers, the contextual stagings of the 'set' – the physical constitution and boundaries of the dancing experience – are vital for their performances within those stagings: their varying inhabitations and constitutions of these stages over the course of the night. Thus, notions of backstage and front stage (density and orientation of the crowd and of the individuals within that crowd, the intensity of the dancing across the differing spaces within the club, the proxemic relationships clubbers develop with those around them), notions of display and concealment (lighting/darkness, visibility/invisibility, smoke/dry ice, lasers, strobe lighting, centre-stage/backstage), and notions of group collusion and teamwork (orientation and communication with the dancing crowd, co-operation and synchronisation with other dancers) are not quite as open and 'up-for-grabs' as they might at first appear. The key scriptings of style and coolness, of personal abilities and learned competencies, and the impact of production and staging – the directing influences of music, of timings, of volume, of lighting (visibility) – are blended imperceptibly in the 'tactical' (re)readings of these scripts and directions, and with the negotiated nature of the dancers' inhabitations of stage, set, props, scenery and costume. Through deliberate and accidental ad libbing, dancers can veer away from prescribed and/or dominant practices, scriptings and spacings in creating unique and meaningful roles of their own.

BEN: [in a joking voice]…so, Seb, do you think you're a 'good' dancer?

SEB: Do I think I'm a good dancer? Oh, let's be blasé about this – I like the way that I dance [do you?]. I definitely like the way that I dance. I think it's kind of an expression of yourself. Your dancing always changes – I mean I remember the way that I used to dance when I first went to clubs. I remember the first time at Fresh, even before the drugs and stuff, watching people dancing and thinking 'My God, they're brilliant! – I couldn't possibly go up on that stage because you've got to be someone to be up there!'. It took me a long time to realise that it doesn't matter what you do or how you look as long as you're having a good time and I didn't accept that and I now accept that other people dance in different ways. I think my sister's a great dancer – some of the things she does I like, and some of the things that she does I don't like. When I see someone dancing I try and mimic what they're doing and make it my own. My mate Matt's a fantastic dancer – he's got that party piece that he does that's just got me in fits. I love the way I dance because it's an expression of who I am – I'm a very right-handed dancer [yeah, you said]. I noticed that especially when I was with Dan – I can take my right hand and use that and only that. Ummm, I can try with my left hand but it never lasts as long as my right. I'm very much a…I try and get people around me involved. I try and single one person out and try and dance with them and if I can get a 'dance chat' going with them then that's brilliant, but I'm very much a stage person, hands in the air, kind of bring people up with me. The night at The Shed when I was just totally off it, that's the kind of time when I really let loose and…one of the reasons why The Shed was such a good night for me was it gave me the chance to dance the likes of which I wouldn't do in Fresh. Because I was on my own and no one there knew me [ummm] and I'm not going to see those people again, I can just totally lose it and do what I want to do and if I think that it's accepted, then it gives me confidence to maybe try it with people that do know me, but that's what I like most. When I'm in a club with people that do know me I think…there's a little bit of holding back, accepting that other people are aware of what you're doing and I think I'm conscious of that.

BEN: Do you often look down at what your feet are doing?

SEB: As a rule I don't move my feet when I'm dancing, not much at all, not more than a shuffle anyway. I don't dance with my feet – I used to when the music was different, but now it's all in my arms. Yes, I'm always conscious of what my arms are doing. There's always times when I'm dancing, especially early on in the night, that I start doing stuff. I might do something new or I might do a move that I know how to do and if it's worked then I'll know that it's worked and I'll know that it's looked good and I try and lose it and just go with the music and let my arms and the music go as one. If it works – brilliant, and if I see it not working then I'll make a little bit more effort.

BEN: What does 'not working' mean?

SEB: Just smoothness…it flows. If I can go from one rhythm to another smoothly, so it fits in. There are different kinds of moves that I've got – hands going up, hands going down, side to side…if I can get from one to another smoothly and it fits in with the music – brilliant, and if I recognise a tune, something I do when a tune is building and building is I gradually bring my hands up as it builds and I can hold it and kick it off at the right time. If I don't know the tune and it's a very long build up and I go up too soon, then I think 'Shit, I've gone off too soon here, I'm just going to hold this and not look like a dick' you know? It's choreographed, that's just the way I am.

In this dialogue Seb begins to evoke a sense of the intense interaction on the dance floor. Seb wants other clubbers to be on the same level as him, to dance with him and to share his particular understanding of the music. He gains a sense of deep pleasure from being seen as, and understanding himself to be, 'good' at dancing; that is, consistent and flowing in his movements. Dancing becomes a source of both social and personal identity. Clubbing experiences where little is at stake, such as the club where he was on his own, that is, without his regular clubbing friends, provide opportunities for experimentation – 'I can just totally lose it and do what I want' – yet he still seeks 'acceptance' from these clubbers despite knowing that he will never see them again. Towards the end of the extract it is possible to discern how the relationship between the micro-practices of dancing and its relationships with the structurings (rhythms, narratives) of music are so important to Seb. He times his 'moves' precisely so that they fit in with the musical rhythms, and if he gets it 'wrong' and goes 'up too soon' then he ad libs his way through his 'error' so that he does 'not look like a dick'. Seb acts within what is in one respect the totalising impact of the music – it dominates the space and the crowd – yet he also improvises and creates his own scripts within the apparently all-encompassing ordering of that music.

With Seb's view firmly in mind, I now want to unpack a conceptualisation of dancing as a set of concurrent spacings. I use the term 'spacing' in place of 'space' for similar reasons that 'consuming', 'ordering' and 'clubbing' are more useful conceptually than 'consumption', 'orders' and 'clubs' respectively (see Hetherington, 1997; Law, 1994). Spacings differ conceptually from spaces in that the former are explicitly 'never finished', always open to negotiation and thus always in a process of becoming. Further, many 'sets' of spacings (or spatial orderings) may co-exist within the same physical space. Each spacing, that is, each understanding of that space, will have its own usually unwritten codes, rules, symbols and customs and will generate its own systems and relations of power (Hetherington, 1997). These rules, customs and rituals form the processes of sociality that may have evolved over time, across spaces and through experiences (which I introduced in Part One). Crucially, these techniques and competency of their practice must be worked at if 'mistakes' are to be avoided.

The spacings or spatial orderings of dancing are never fixed or finished then, but are continually in process. They are not singular, although there may be a dominant ethos or understanding at any one moment or in a certain area, but rather multiple and overlapping. Different clubbers will have different spacings of clubbing, and these will vary over times and between spaces. Conceptually, and with each differing in scale, the spacings of dancing are constituted in at least four

important and inter-dependent ways, namely: territorialisations and regionalisations; mediations; techniques and competencies; and emotions. I deal briefly with
each of these in turn, although of course it should be stressed that they constitute
the clubbing experience concurrently.

Territorialisations and regionalisations

Dancing is constituted firstly through the territorialisations and regionalisations
that its practices inhabit. The dance floors, the different areas within those dance
floors, notions of front stage and backstage, areas for sitting and drinking that
surround them, cloakrooms, loos, eating areas – are all territorialisations that are
bounded both spatially and temporally and are always open to alternative spacings
and orderings that dancing and non-dancing clubbers may attempt to impose[33]. In
time these alternative orderings may become momentarily dominant themselves.
Different practices are 'acceptable' within the differing regionalisations of the
clubbing space. While dancing on the dance floor will usually be entirely acceptable, dancing at the bar, where the 'rules' of tactility and bodily understandings of
music are quite different, can project a lack of tact and even a loss of bodily
control; one risks displaying a lack of belonging. Drinking at the bar is completely
normal, whereas drinking in the middle of the dance floor can convey a gross lack
of awareness and drunkenness and is certainly uncool. Failure to recognise the
boundaries of these regionalisations can signify incompetence, with special significance given to the dance floor perimeters. These perimeters might be physically
established, for example through the provision of a clearly demarcated, often
wooden, dance floor. In any case these boundaries will vary according to the
particular stage of the evening; as the climax of the night approaches, for
example, it is possible that the floor area of the whole club might effectively
become dance floor.

Differing regionalisations of the club provide greater potential for 'involvement shields' (Goffman, 1963: 38); that is, opportunities to momentarily come
'off-stage' or 'go out of play' (ibid.: 40) and interact with others in a much less
active fashion. Club loos, the bar, cloakrooms, inter-dance-floor corridors are all
forms of 'involvement shields' or forms of 'backstage' regionalisation, yet so – at
times – is the darkness, semi-visibility and sensory onslaught of the dance floor. It
is possible to 'disappear' in(to) the crowd. Competency at understanding the
regionalisations of club spaces through a largely self-imposed self-education
allows the dancing clubber to negotiate these complex socio-spatial demands. The
spacings of the club are thus also bounded in the sense that they are constituted
through technical competencies and a knowledge of 'local' customs, rituals and
modes of behaviour. Different areas of the club are used by the clubbers for
various activities at different times of the night. The dance floor itself is often

further sub-divided by the dancers, with a common notion being that of a 'personal realm' – a notion of 'a space of our own' (a specific place for themselves and their friends) – which is more or less colonised throughout the night. These regionalisations can change from night to night, as well as during the course of a single night out.

BEN: Are you conscious of any differences that you get in various areas of the club?

LUKE: As a matter of fact, most of the time I move within a relatively small area of a club. Every place in a club seems to attract a different type of people, even in small clubs. Sometimes it even goes so far that I can feel a little uncomfortable when being in the 'wrong' area. Not that I am afraid of anyone or anything, it's just that things don't seem to fit...light, music volume, people around you...on the other hand, it can give sort of a new experience to 'explore' a new area sometimes. On an average night, I spend about two thirds on my own spot, by which I mean an area of the club, including some space on the dance floor and some space around the bar and seats, and one third 'being on vacation' to different areas. Now that I think about it, I realise that 'my area' is almost every time close to the speakers. I can tell you another story about a rave where I experienced the different types in a very extreme way. The main room was very large and in the middle there was a stage in the form of a cross. The good thing about that was that every corner of the cross acted like a little area, though in fact everybody was in the same room. I spent most of the night in just one corner, because while walking around, we saw that another corner was the place to pull up, that one corner turned out to be the home base of the Essex-like guys and the other was really nothing. And of course, I was in the corner with the nice, friendly and open-minded people!

Luke is outlining how and why he uses different areas of the club through the night. His description of an area in which he stays most of the night is in common with those of other clubbers to whom I talked, with many speaking of a 'space' or 'territory' – Luke calls it 'his area' and 'his spot' – which they 'colonise' for the night. This has the role of providing immediate security in the face of huge numbers of people, as well as acting as a meeting space for the clubbing group that the clubbers have either met or with whom they came. The varying nature of different areas of the dance floor and the contextually specific practices deployed there are indicated by Luke. For example, certain areas are better for 'pulling', while other areas are where the groups of lads hang out.

Mediations

A second set of spacings that constitute the experience of dancing are mediations – the various environmental effects and qualities that exist within, and in many cases assist in defining, regionalisations and territorialisations. I outlined earlier how a central mediation is music which, even within a single club night, can be spatially variegated in terms of different areas and rooms for different musics, and temporally variegated according to the stage of the evening and the particular DJ in control 'on the decks'. A change in music can radically change the constitution and expressive form of a dancing crowd, and the use of 'anthemic' and 'classic' tunes plays an important role in transforming the crowd's intensity and numerical constitution.

Supplementing the music are the environmental effects of lighting (or lack of it), thick smoke, dry ice and even 'raining foam', acting both to sensuously disorientate and physically insulate the clubbers from, and within, the surrounding crowd, even if only fleetingly. The combination of these effects, together with the extremely loud music, can evoke the sensation of being in a slightly unreal world, a space in which there is almost too much sensuous stimulation at once. Furthermore, the clothed, moving and interacting bodies of the often closely packed dancing crowd are themselves adding to this aura of unreality and 'other worldly-ness'. With clever manipulation of lighting and the targeted use of visual effects, such as slide show loops and film projections – both further 'directing' influences of production – an illusion can be fostered of being in an-other place at an-other time, of momentarily inhabiting a dream-world, of being beyond or outside 'normal' time and space.

Techniques and competencies

Third, dancing is constituted and experienced through the dancing clubbers' techniques, competencies and spacings. Clubbers develop techniques and competencies – in the form of sociality – in the use of their bodies in such a manner as to successfully negotiate the trials of 'impression management' which present themselves across the various and changing regionalisations of the club in the form of sociality (Goffman, 1959). This impression management can be an individual matter, but more usually involves 'team collusion'. Thus, the successful deployment of techniques on the dance floor involves co-dancers who to some extent share access to a wide range of information about these techniques. They agree on what should be understated and suppressed (what does not look good), as well as agreeing on what should be actively presented (what does look good) and on basics, such as the physical division of dancing space.

Perhaps the most obvious, most significant forms of body techniques[34] (Mauss,

1979) involved in dancing are the *spacings* of their co-present bodies. The dancers co-operate in distributing themselves throughout the available and constantly changing space on the dance floor to provide themselves with maximum personal space, yet at the same time also to facilitate 'engagement closure' through this spacing (Goffman, 1963: 156)[35]. The nature of these spacings will change continually throughout the night as the dancing crowd changes in size, intensity, constitution and levels of drug consumption.

Added to these bodily spacings are the techniques of engagement with co-dancers in the dancing crowd. Drifting between consciousness of their self and consciousness of the crowd, clubbers also vacillate between dancing 'alone' (dancing with themselves) and dancing 'with others', as part of and responsive to the dancing crowd. Some clubbers, such as Seb, preferred the sense of freedom that they experienced through dancing alone in various areas of the club (at least at times). Other clubbers, such as Valerie, tended to remain together within the group with whom they came for the duration of the night. Still others, such as Luke, said how they enjoyed the sensation of being *between* the crowd and the sense of aloneness that the crowded dance floor could instil.

In terms of dancing styles and the specific bodily practices of the clubbers, it is evident that different clubbers have different ways of dancing even where one dancer attempts to emulate another. An individual clubber will dance quite differently according to the stage of the night, the type of night, whether they have consumed drugs or alcohol, the area of the club in which they are, the nature of the immediately proximate clubbing crowd, the music and their level of competency. There are, nevertheless, very well-established 'ways of doing' dancing, in a similar way that I suggested earlier in 'The night out' that there were specific skills at which one had to attain competency to successfully negotiate entry into the club. Dancing is one way of speaking through the body both to others and oneself, and as such when people dance they usually like to be saying the right things, or at least be seen to be in control of what they are saying. Not only have videos specifically on 'how to dance' been released and re-released (although not surprisingly I have yet to find anyone who has bought one), clubbing magazines regularly feature articles on dancing and dancing techniques. These articles are often presented in an ironic and somewhat mocking way – again a reflection of the discomfort that most clubbers experience when talking about their own dancing abilities, yet, at the same time, they do usually contain relatively accurate diagrams and descriptions of dancing clubbers (see, for example, Cook, 1997).

Appearances are an important facet of sociality in themselves – we come to believe that we are momentarily seeing into someone's personality when we see someone dance – and the broad sharing of dancing styles can be a powerful catalyst for development of (micro-)identifications, as was evident in the extracts from the interviews with Luke and Seb. In a similar way to dress and other forms of bodily adornment, dancing provides a way for the individual to present his or her self-image to others, as well as to his or her self (Finkelstein, 1991). Supplementing the more immediately apparent techniques and spacings of body movements are the less obvious body techniques of gazings, touchings, gesturings and other forms of sociality that comprise 'dancing'. Managing one's body requires practice. As Butler (1990a) infers, the dancing body becomes a variable boundary, a style of being through which social, and thus self-identity, is not only transformed, but actually becomes possible in the first place.

Clubbers learn how to dance through a combination of watching others (mimicry) and through listening to and understanding the music. The exact form of their 'education' takes on differing forms depending, for example, upon the genre of clubbing concerned, the age and gender of the clubber, the clubber's past experiences and the clubber's view of his or her own relationship to the clubbing crowd. In its combination of listening, mimicry and embodiment, dancing represents a more spatially constituted form of listening to music than, say, watching a band perform on a bandstand or watching an orchestra play in a concert hall because understandings are more actively traced and visibly performed in space. Through their actions in inhabiting and constituting spaces, dancing clubbers express a listening experience. Through the techniques of the body in understanding music they can explicitly identify with others through mimicry, with a style, through the display or emulation of that style, and with a crowd through combining proximity and tactility with this mimicry and its style.

It should be re-stated that the body techniques of dancing were, of the four sets of spacings I am conceptualising, the spacing about which the clubbers were the least happy, or perhaps least able, to talk during the interviews. Self-evaluation, when it did occur, tended to be restricted to value judgements about whether another person was dancing 'well' or not, or was ironically self-deprecatory in nature. However, I agree with one of the clubbers, Sun, in believing that this 'reticence' has more to do with the difficulty of *talking about dancing* than with any lack of significance of dancing to the practices of clubbing.

Emotional spacings

A fourth set of spacings, and the most elusive to pin down both practically and conceptually, are the emotional spacings of clubbing. These arise through the complex interaction of territorialisations, mediations (physical effects) and bodily practices, with notions of identities and identifications. These emotional spacings are particularly significant to the dancing experience in that dancing is at once both a cause and an effect. Emotions are expressed through the body, yet concurrently instilled through bodily movement, in tandem with music and other mediations.

The emotional spacings of dancing are the most personal and 'local' – the most 'up-for-grabs' and pliable – of the four forms of spatial orderings that I am conceptualising. As clubbers become more competent at the practices constituting each set of spacings, they gain confidence at addressing increasingly personal, and thus more fluid, problems of competency and can 'travel' towards these usually pleasurable emotional experiences of dancing.

Conceptually decreasing in scale from 'environment' to 'self', clubbers can develop proficiencies at understanding and acting within the territorialisations and regionalisations of the various areas (or regimes) of the physical club space as a whole (notions of inside and outside, spaces of dancing and non-dancing, of expression and restraint, of involvement and of non-involvement). They can develop a level of confidence and competency at knowing and enjoying the various mediations of music (its volume, speed [bpm], beats, rhythms), of lighting and sensory stimulation produced through 'directing' (what to do where and in what way) and through the bodily practices of the dancing itself (how to move arms, legs, head, fingers, feet; when to open or close eyes; in what ways to embody the rhythm; how to express happiness and euphoria; how to interact through dancing; how to deal and interact with others who are extremely close; if, where, and how to gaze and be gazed upon).

As competency at negotiating each of these spacings is developed, the clubbers move towards addressing and understanding dancing through emotional spacings. The challenge of mastering – or for some merely coping with – the spacings of regionalisations, mediations and bodily practices at once is such that a complete engagement with the emotional spacings of dancing is rare and when it does occur is usually fleeting. The dancer is usually at least partially monitoring other facets of the dancing experience. If successful in meeting the spacings of regionalisations, mediations and bodily techniques through understanding where and how to use one's body, a sensation of momentary introspection (or *exstasis*) may be experienced.

During these sometimes deeply emotional moments – and 'moments' are usually quite literally their duration unless drugs are involved (see p. 119) – the individual clubber is outwardly participating in the practices of dancing, but inwardly momentarily allows his or her attention to switch from this dancing to a play-like and reflexive world in which s/he alone participates. These relationships between bodily actions and emotional and mental reflection can be forcefully experienced – 'dance is a powerful, frequently adopted symbol of the way people feel about themselves' (Royce, 1977: 163). Goffman (1963: 69) usefully describes this mixture of detachment within togetherness as an 'away', and this term effectively and concisely evokes the sensation experienced by stressing the retention of a close control over one's body, yet a simultaneous sense of situational detachment. As the dancing clubber fleetingly experiences this 'away' and fluxes from

THE DANCER FROM THE DANCE

consciousness of environment to that of self, the 'directing' influences of club production and staging (the territorialisations and mediations of production) become less or even insignificant as these spacings (and their scriptings) are progressively more open to deliberate ad libbing by the individual dancing clubbers. In this way dancing offers to both individual clubbers and the group of which they are a part 'the chance to become something else "for the time being" or...to rebecome what they never were' (Schechner, 1981: 3)[36].

ROGER: I danced for about two-and-a-half hours altogether, a bit more than average. I love the mixture of community and isolation on the dance floor...you can go for ten minutes with your eyes closed, caught up in the mechanics, then look around and catch people's eyes and exchange grins. Sometimes I lay down some tasty moves...although God knows what it looks like, but who cares! At other times I just do kind of the same movements over and over and think or daydream or just look around. Sometimes I dance 'with' people, sometimes on my own...it feels different and yet the same every time. I always start off feeling like Nelly the Elephant and two songs on I feel I have become Nureyev...I wish!

Roger's brief account of dancing is noteworthy for at least five reasons. First, the isolation–community tension is evocative of the in-betweeness that I have been stressing throughout 'The night out' thus far. Second, his notion of being 'caught up in the mechanics' chimes with the notion of loss of bodily control that I turn to on p. 135. Third, and another facet of this in-betweeness, Roger fluxes between dancing with himself and dancing with others, while he is always, of course, within the dancing crowd. Fourth, through feeling 'different and yet the same every time', dancing provides both novelty and familiarity. Finally, Roger is quite aware of what 'being good' at dancing means and can be openly self-deprecating about his abilities, as his comparison of Nelly and Nureyev demonstrates.

I turn in detail to the moments of introversion to which Roger alludes on p. 105 – moments that are of immeasurable importance to the popularity of the clubbing experience, and which I describe as 'moments of ecstasy' or 'oceanic experiences' – in the following two sections of 'The night out'. I want to complete this section of 'The night out', and thus also this initial discussion of dancing and music, by noting a final ordering influence that affects and constitutes each of the spacings, which I have just outlined, in a significant manner. This ordering influence is temporality.

Temporality

Due to licensing restrictions, club nights almost always start and finish at set times, so the spacings I have outlined are primarily defined temporally in that the clubbing experience lasts for a specified period. During the experience, the presence of music, and especially of rhythm, powerfully impacts upon the clubbers' notions of time as well as their experience of spaces. While the music is playing, the notion of linear time passing can seem to have changed. Indeed, as Blacking points out, one of the essential qualities of music is its power to seemingly create another world of virtual time: 'We often experience greater intensity of living when our normal time values are upset...music may help to generate such experiences' (Blacking, 1973: 51)[37]. The massively loud volume of the music typical in most clubbing experiences not only virtually obliterates verbal communication, especially on the dance floor, but it can also act to structure time for those present, while they are present, as eternally *in the present*, as a momentary time of continual 'now-ness'. Through imposing sonic orderings and spacings upon the social gathering, music can affect emotional responses and can in certain instances effect a coincidence of emotional arousal at the same moment (Storr, 1992).

To dance is not just to experience music as time, then, but also to experience time as music, to experience a set of spacings that are cut off or divorced from everyday time (Frith, 1996). Time passing is superseded by music playing. Time and its intense regulation of bodily actions *outside* clubbing – through the demands of working, of travelling, of general sociability and interacting – is replaced by music and its regulation of bodily action *within* the spacings of clubbing – through dancing, through spatial selection, through bodily expression of emotional and identificatory responses to the crowd.

Dancing clubbers often talk about forgetting time. The clubbing experience effectively transforms time into space and music for the clubbers whilst they are clubbing. The 'now-ness' of clubbing, its insistence on the present and the significance of the moment repeatedly reinforces the notion of clubbing as somehow outside time, regulated temporally only through rhythm. The focus for the clubbers is on *what is* rather than *what should be* (Maffesoli, 1987), and the pleasures of the present may temporarily obliterate the concerns of life before, beyond and outside the clubbing experience. To give an example, a very common feature of the end of a club night is the palpable sense of disbelief that the night might actually be over, despite the inevitability of this moment. The fluorescent lights coming on overhead not only reveal the usually (by now at any rate) less-than-beautiful bodies, they also indicate that 'time' is back, that the experience is effectively over, that this space only exists within the wider city waiting beyond the door.

This 'succession of presents' (Maffesoli, 1995: 145) that is characteristic of the ambience of clubbing further contributes to an ethos in which proximity to and identification with others is foregrounded, but in which there is also possible a strong sense of introspection. Clubbers are both submerged beneath, and yet are relatively easily able to fall in and out of, the clubbing crowd.

> Music is not, by its nature, rational or analytic; it offers us not argument but experience, and for a moment – for moments – that experience involves *ideal time*, an ideal defined by the integration of what is routinely kept separate – the individual and the social, the mind and the body, change and stillness, the different and the same, already past and still to come, desire and fulfilment.
>
> (Frith, 1996: 157; emphasis in original)

OUT OF SPACE, OUT OF TIME

Clubbers' understandings of the rhythms, beats and lyrics of music are central to their enjoyment of the clubbing experience. Music is central to the practices and spacings of clubbing. Not only can music intensify shared experiences through exaggerating, and in some cases distilling, emotional responses brought forth by that shared experience, but music can alter regionalisations to such an extent that it prevents verbal interaction between those sharing the experience of that space. Music can be ascribed a totemic role, and can act as a 'mainspring...which binds people together as well as liberates them' (Maffesoli, 1995: 144). Music can act as a focus for the articulation of identities and identifications.

Music is also central to the practices and spacings of dancing – practices and spacings which can be understood as embodied understandings of that music and as a form of creative listening that prioritises these shared knowledges of music and its embodied performances. For clubbers, dancing is about simultaneously losing yet also gaining control over one's body; dancing is about becoming part of and submitting to the dancing crowd. Clubbing crowds are at once both emotional and motional, both social and spatial formations, both expressive of self yet also constructive and transformative of self. Momentarily displaced, lost between the community and the isolation of the dancing crowd, the clubbers become as unified and inseparable as the dancer and the dance.

THE NIGHT OUT

O chestnut tree, great-rooted blossomer,
Are you the leaf, the blossom or the bole?
O body swayed to music, O brightening glance,
How can we know the dancer from the dance?

(W. B. Yeats, 1956)

Out of space, out of time,
Out of sight, out of mind,
Feeling you feeling me,
Feeling me feeling you –
We're flying, together,
We're flying in each other's arms.

(Marlena Shaw, 1977, 'Look at me, look at you')

MOMENTS OF ECSTASY

Oceanic and ecstatic experiences in clubbing

That unmatched form and
figure of blown youth
Blasted with ecstasy.

(*Hamlet*, Act III, scene i, line 170–71)

INTRODUCTION

The dance floor is now bustling. Sound and movement combine to give the impression of chaos. A pleasant experience of sensory overload can occur. It is dark, crowded, very warm and humid. The lighting appears to be linked to the music, which is now penetrating minds *and* bodies, going through rather than around clubbers as its intensity increases. Many clubbers appear to be almost picked up by the music and commanded or orchestrated in their dancing, unable to stop, not wanting to stop, unable to even conceive of stopping. This experience is characterised by moments of extraordinary euphoria and happiness.

In this third stage of 'The night out' I open out a conception of two closely related but significantly different forms of what are conventionally labelled 'altered states'. I argue that these 'altered states' characterise, and for some clubbers provide the *raison d'être* of, the clubbing experience. This section of 'The night out' is split into three sections. First, in outlining the notion of what I am calling 'oceanic experiences', I examine in detail these sensations of extraordinary and transitory euphoria, joy and empathy that can be experienced as a result of the intensive sensory stimulation of the dance floor. Notions of freedom and of in-betweeness appear to be vital, yet these are melded with the strong sensations of belonging and identification that I introduced earlier.

Second, I discuss a major form of oceanic experience – the 'ecstatic experience' – which is premised upon the use of so-called 'dance' or 'recreational' drugs such as ecstasy (MDMA). The use of ecstasy (MDMA) in particular appears symptomatic of many clubbers' quest for the fleeting moments of complete contentment that it is possible to experience while clubbing, particularly within the dancing crowd.

Third, I build upon these points regarding the ecstatic experience and, in tracing out 'A night on E' in some detail, I further develop an understanding of the often quite technical practices, spacings and timings of these 'moments of ecstasy' which, for many, are the very essence of clubbing. Far from representing a shallow and meaningless, simply hedonistic and purely selfish experience, I argue that the oceanic experience and its drug-assisted variant can provide powerful sensations of personal and group identity formation, amendment and consolidation.

THE OCEANIC EXPERIENCE

Of all the arts, music is undoubtedly the one that has the greatest capacity to move us, and the emotion it arouses can reach overwhelming proportions.

(Rouget, 1985: 316)

Although it usually takes place within a social context, the dancing that is so central to clubbing is a highly personal experience and yet one that is rarely evoked in writing because of its ineffable and ephemeral qualities[38]. The contrasting moments of introversion and of complete engagement with the dancing crowd that the dancing clubber may experience, upon which I touched at the end of the last section of 'The night out', are impossible to convey fully through words alone. In my attempts to evoke these experiences, which I am variously conceptualising as 'oceanic' and 'ecstatic' for reasons I outline on p.109, it is thus important that the frames of reference and terminology that are used are explained as fully as is possible at the outset. While Lewis (1989: 10), for example, is quite correct to critique over-elaborate and fanciful concepts of altered states and ecstatic (and other so-called 'mystical') experiences as flawed and based upon 'elaborate structuralist dichotomies', he firstly fails to outline any alternative, and secondly he partially falls into his own trap by embarking upon a lengthy and confusing chapter of debate over definitions. When discussing these most individual, fleeting and powerfully emotional experiences some clarification of terms and concepts is necessary and should not be avoided simply because of their complexity.

At the risk of tautology, at their most basic level of conception 'altered' states of mind differ from 'normal' states of mind in being qualitatively unlike the predominant states of mind experienced during one's waking hours in the course of 'normal' day-to-day living. They involve some form of transformation in consciousness (Inglis, 1989; Laski, 1961; Lewis, 1989). Thus, 'altered' states do not include common experiences such as euphoria, happiness or joy as experienced on an *everyday* basis, but rather only euphoria, happiness and joy characterised by a transitory, unexpected, valued and *extraordinary quality* of rare occurrence and magnitude in which an altered sense of consciousness is temporarily experienced (Laski, 1961; 1980).

Throughout 'The night out' I have referred to the experience of in-betweeness or liminality – of being somehow taken outside of or beyond oneself, especially while dancing – that is a characteristic of many crowds and particularly of the closely packed, sensorially bombarded dancing crowds of clubbing. At its most intensive this sensation of in-betweeness can partially induce or trigger the altered state of consciousness that I am calling, after Freud (1961) and Storr (1992), the 'oceanic' experience. I use this term in evoking 'a feeling of an indissoluble bond, of being one with the external world as a whole' (Freud, 1961: 65).[39]

Drawing from work by Marghanita Laski (1961; 1980) on 'ecstasies', by Brian Inglis (1989) on altered states such as 'trance', by Anthony Storr (1992) on music and the emotions, and by I.M. Lewis (1989) on 'ecstasies' and altered states more generally, I define as 'oceanic' those experiences characterised by one or more of these sensations: ecstasy, joy, euphoria, ephemerality, empathy, alterity (a sense of being beyond the everyday), release, the loss and subsequent gaining of control, and notions of escape[40]. In her work on altered states, Marghanita Laski describes an oceanic experience (or what she calls an 'ecstatic experience') 'as being one in which all sense of self and time and the everyday world seem to vanish...a state of anxiety is replaced by mental tranquillity' (Laski, 1980: 12–13). Pleasurable fluctuations between awareness of self and environment, between sensations of intensity and withdrawal, and between practices of interaction and reflection are foregrounded.

For Laski, oceanic experiences are usually described as indescribable, although this does not prevent people trying to describe them in some detail – they just usually never quite 'get there'. Laski also suggests that oceanic experiences might involve feelings of loss (of self, of time, of place, of limitations); feelings of gain (of unity, of 'everythingness', of oneness, of an ideal place, of release); and feelings of 'quasi-physicality' – of some form of discontinuity between the physical and emotional experience of one's own body and surroundings (Laski, 1980: 14). People describing oceanic experiences often evoke their feelings by describing sensations of upness, swelling, warm flushes or glowing in the heart, feelings of warmth, of liquidity, tinglings in the head and spine, and the attainment of calm and peace (Laski, 1961; 1980). Happold (1981) appears to concur with Laski in proposing that 'mystical states' (as he calls oceanic experiences) are marked by ineffability, transience and passivity, and often instil a sensation of the oneness of everything.

Laski (1961) lists three main and prior uses of the term 'ecstasy' in addition to her own use as a form of oceanic experience: first, in describing a trance-like state; second, in describing a state of madness, as in Ophelia's description of Hamlet as being 'blasted with ecstasy' with which I opened this section of 'The night out'; and third, in describing a state of being in love, particularly, Laski suggests, in advertising media. However, because of current connotations with drug consumption this latter use of the term 'ecstasy' – in advertising media – is now highly problematic and would probably be subject to some form of Advertising Standards Authority control. Of course, only two years after Laski had written her second text on ecstasy, the term had taken on a fourth and additional connotation of referring to a 'hallucinogenic amphetamine' (ISDD, 1996: 2; Stevens, 1993).

In contrast to the more usual considerations of the oceanic experience – where solitude appears to be a pre-condition and feelings of unity are experienced more with a 'god' or 'universe figure' than with any 'earthly' manifestation or 'thing' – I am using the concept to evoke the sensation of oneness and the liminality of self/wider group that can be experienced within apparently diverse dancing crowds (Storr, 1992). These experiences can induce highly pleasurable moments of *exstasis* (or loss of self) within individuals.

The notion of the oceanic experience is appropriate in discussing the altered states of consciousness that clubbers describe as sometimes experienced while dancing in crowds for at least four reasons. First, as I have suggested, the oceanic neatly evokes the sense of in-betweeness or liminality that characterises clubbing and particularly the practices and emotions of dancing: in-between spaces (outside/inside, inner night-life/outside everyday life, spaces of work/spaces of play) and in-between times (night/day, work/play, and even outside time, or between 'real' times). Second, the oceanic evokes the fluidity and constantly shifting socio-spatial dynamics of the dance floor – the sensory onslaught that can act to effectively remove any figure of reference (the walls merge into the darkness, the ceiling is invisible, the floor is rarely glimpsed) – and the unceasing motion of the dancers. Third, as both Storr (1992) and Lewis (1989) note, music alone rarely triggers the oceanic experience. Yet, through dancing and the embodiment of that music, through self-mastery and the use of body techniques in the expression of a dancer's understandings, the dancer is able to transcend or escape the self and strive for a realm beyond the confines of the body – 'to move one's body is to aim at things through it' (Merleau-Ponty, 1962: 137). Fourth, and related to these first three points, the oceanic experiences of the clubbers to whom I have talked appear to foreground (as does Laski's concept of the ecstatic) notions of *loss* (of differences between self and others, of time and space, of words, images and the senses), as well as notions of *gain* (of unity, of timelessness and eternity, of control, joy, contact and ineffability).

I refer to oceanic experiences that are partially attained through the use of chemical triggers, in particular the dance drug ecstasy (MDMA), or similar hallucinogenic amphetamines (ISDD, 1996), as 'ecstatic', and I discuss these further on p. 116. The use of the term 'ecstatic experiences' to evoke the broad range of experiences of in-betweeness, *exstasis* and joy that clubbers have talked about both whilst dancing and off the dance floor would perhaps have been less confusing. However, as I mentioned earlier, in an attempt to prevent *all* such experiences being associated with the use of ecstasy (MDMA) or other drugs (a common and rather lazy misconception), I will refer only to those experiences that *do* involve drugs as being 'ecstatic' experiences. The term 'oceanic experiences' is thus used to encompass *both* drug induced (ecstatic) and non-drug induced sensations of in-betweeness. Simultaneous feelings of disassociation and of warmth and empathy towards others – sensations of introversion and meditation yet concurrent expression and *jouissance* – are facets of all crowd-based 'oceanic experiences'. 'Ecstatic experiences', on the other hand, at least in the sense that I am using the term, are those oceanic experiences in which drugs, and particularly in the clubbing experience the drug ecstasy (MDMA), are used in an attempt to trigger, prolong or intensify the experience.

The nature of the oceanic experience in clubbing

I don't know how to put it into words – forgetting oneself, no, oneself ceasing to matter and no longer being connected with everyday things, with the commercial sort of life one lives – a feeling that for the first time you're seeing things in proper proportion – you know that the things in the women's magazines aren't worth anything compared with the leaves on the trees, say, – and time seems to stop, no, not matter, you're not anywhere, and despite not feeling anywhere in particular, feeling in unity with everything – no, not with everything, with nature, but not specifically trees, flowers, plants, everything that comes out of nature, like you might say a book was written by a man, but it's still nature – it feels to a certain extent like a great climax which has built up – this thing has been seething inside you and suddenly it comes out.

(Laski, 1961: 387)[41]

I don't know what the colour scheme is exactly, for it is continually changing. This is caused by ever-varying kaleidoscopic shafts of light – now purple, now blue, now orange, now something else – being projected in turn upon the floor from above. Everywhere light and life and colour and swift movement; the rhythmic tapping of feet; the gleam of backs as bare as they know how to be; the

rustling of attenuated skirts; the throbbing of violins; and, again, the hum of
laughter and talk.

(Wyndham and St. J. George, 1926: 49)[42]

Before I turn to address the use of drugs such as ecstasy (MDMA) in the attaining
of oceanic experiences, I want to note first a number of more general qualities of
the oceanic that can be experienced without (as well as through) the use of drugs,
albeit perhaps in a more fleeting fashion. Moments of oceanic contemplation in
clubbing are characterised by notions of membership of the clubbing crowd, by
their momentary nature, by a tension between intensity and withdrawal, and by
their technical and thus contextual quality. I now briefly address each of these
points.

The oceanic and crowds

In her seminal study on the experience of ecstasy, Marghanita Laski (1961: 177)
proposes that the crowd is an example of what she calls an 'anti-trigger', an
inhibitor to the experience of the oceanic. While recognising that the crowd can
be a source of real pleasure for those within it, Laski suggests that because the
crowd is such an ordinary, everyday aspect of 'normal life' — at least in cities we
experience crowds on a routine, daily basis — it can impede a full sense of loss of
self and reflection[43]. Other commentators see the crowd context as centrally
implicated in the experience of the oceanic. Lewis (1989), Gowan (1975) and
Csikszentmihalyi (1975a) all infer that the experience of the oceanic may also be
marked by an involvement with the crowd, even if, as Lewis (1989) adds, individual experiences within that crowd will have a personal and unique quality to
them. In contrast to Laski (1961; 1980) and as I intimated earlier, I am proposing
that the oceanic, attained with or without the use of drugs, is not dependent
upon, nor induced only through solitude. The oceanic can also be experienced,
albeit in a differing form, in the dancing crowd — a crowd certainly not experienced on an everyday basis. With the exception of the use of drugs, the contexts
of the dancing crowd and the loud music are, in fact, the two most significant
features of the dancer's route into experiencing the oceanic.

Oceanic moments

The oceanic as experienced by those *not* using drugs almost invariably lasts 'an
immeasurably small space of time: an instant, a moment' (Laski, 1980: 14).
Nevertheless, this does not prevent the moment of the oceanic often being characterised by sensations of timelessness, of time temporarily having no meaning, as
I suggested earlier in my discussion of dancing. This ephemerality or timelessness

is due in part to the immense difficulty of suspending the conventional notions of civility, propriety and sociability – notions of consciousness of self and one's social obligations – for anything more than fleeting moments. In addition to the 'enculturated' monitoring and consciousness 'systems' that tend to retrieve dancing clubbers from their reverie almost as soon as they find themselves 'there', there is the more prosaic need not to hit another dancer in the face with a flailing arm. Self-consciousness and a sense of location return almost as quickly as they melt away and the 'default' settings of everyday life – the timings, spacings and monitoring of the body and its presentation and the monitoring of others' bodies and their self-presentation – again re-assert themselves.

However, while the oceanic is usually described in terms of instants or moments, there is often a period of 'afterglow' that persists within the individual after the experience (Laski, 1961: 59). This afterglow can prolong the sense of the oceanic having been experienced, if not the actual experience itself. The pleasurable experience of these oceanic moments thus often endures well after their temporal and far beyond their spatial epicentres[44].

Intensity and withdrawal

Experiences of the oceanic are clearly not identical, differing between instances even for the same individual. Laski (1961: 54–5) distinguishes two main forms of 'the ecstatic' (which, I remind you, corresponds to my notion of the oceanic): 'intensity ecstasies' and 'withdrawal ecstasies'. Laski argues that an experience that is predominantly characterised by intensity may be euphoric, intoxicating and elatory, whereas an experience in which a sense of withdrawal dominates (while still an oceanic experience) might be characterised much more by reflection, contemplation and interpretation. Both are usually intensely pleasurable.

Of course, these are only extremes – most oceanic experiences feature both moments of intensity and of withdrawal, both effervescence and introspection. The sensory overload that can characterise a state of euphoria can, through obliterating other aspects of the situation, such as gazings, speech and other aspects of situational interaction, focus the mind within, as well as overwhelm it from without. If the intensity of the situation is such that the dancer is momentarily relieved of the obligations and technical requirements of social interaction, then inner reflection and contemplation may become possible. Through focusing attention and concentrating the senses, music and the other mediations of clubbing can limit the stimulus field, distractions are temporarily eliminated and a sense of merging with the music – a loss of self or sensation of *exstasis* – can be experienced (Csikszentmihalyi, 1975b).

By no means do all clubbers, dancing or otherwise, experience oceanic sensations of intensity and/or withdrawal whilst clubbing. Given the difficulty of articulating these experiences for the clubbers, particularly to someone who is not a close friend, it is not possible to propose with any degree of certainty how many of the eighteen clubbers that assisted me in interviews and clubbing nights out had experienced oceanic sensations[45]. However, all the clubbers attempted to evoke a 'special feeling' that they experienced at certain times while dancing. Some suggested that they were most aware of the intensity of the dance floor, that they were enjoying the massive sensory stimulation of music, closely packed bodies and semi-visibility. Others talked about dancing as providing a time to think, as allowing them to 'lose it' (as Seb suggested during my discussion of dancing in the last section). Still others (such as Roger, p. 101) mentioned both the 'community and the isolation' of the dance floor. The nature of clubbers' experiences of intensity and/or withdrawal – and thus of the oceanic – while dancing appeared to depend, to some extent at least, upon how confident and competent they felt while dancing. The techniques of the dancing body, which I discussed in the previous section as one of the key 'spacings of clubbing', are thus centrally implicated in the emotional and sensorial experience of the dance floor, and in the nature of the emotional spacings that I also mentioned.

Techniques of losing and gaining control

BEN: What were you thinking?

KIM: Ummm...in some ways I sort of switch off if you like, and not think, just...I prefer...the way I'm like responding or dancing to the music or anything. I try...I don't like to think about that too much because if you start thinking about what you're doing or you suddenly think 'oh, got that' and you try and...so like I just try to tend to let me brain...and saying about thinking about my CV or whether I should get my hair cut or something like that, thoughts like that often pop into my head and it's...it does sound really trivial but...one thing that I was thinking about was, and it was quite possibly because you were there and you were doing this interview and stuff, I was thinking about umm...the rave culture and the dance culture and everything else, and why everybody did it etcetera and I was thinking you know, like the video games, the quality of material, everything is so like, lighting, TV...we're getting so much stimulation, really strongly, all the time, that if you want to go out and get...it's going to be that hard to get...and about care in the community as far as giving people things to, well, giving people the opportunity to go out and enjoy themselves and all the things against it and for it...that sort of stuff.

Kim is clearly attempting to relate the experience of the dance floor at that moment to her everyday life and to life more generally – partially, it would appear, because of my presence. She also appears to be struggling to find the words that precisely evoke her emotions and thoughts at that moment. Apparently, Kim does not like to, nor need to, 'think' about her dancing movements, instead using the space and time of the dance floor to think about other aspects of her life. This flux between participation and introversion is common.

In the clubbing crowd, the oceanic, if experienced at all, is usually experienced through participation within the dancing crowd. Although Mauss suggests that 'underlying all our mystic states there are corporeal techniques' (1979: 336), it is wrong to reduce the oceanic merely to the outcome of a correctly practised set of body movements in a specific social context (Rouget, 1985). As with many other aspects of clubbing, issues of competence are not as clear-cut as merely involving the practice of learnt body techniques. For one thing, notions of coolness can also affect the way – the spacings and timings – that a dancer dances. Furthermore, the timings and spacings of clubbing mean that at certain times it is cool *not to succumb* to the crowd and music and actually to resist an available route into the oceanic, perhaps by *not* dancing. At other times the music, crowd and 'e/motion' may become so intense that even experienced clubbers, regarded as 'cool' (if only by themselves), may appear incapable of resisting the summons of the music and the crowd, and thus they lose themselves to the intensity of the situation (Rouget, 1985). As Crossley states: 'to have acquired a body technique is precisely to be able to adapt and apply it in accordance with the demands of particular situations' (1995: 137). Clubbers have differing understandings of when and where it is appropriate to be seen to momentarily surrender control of one's body in experiencing the oceanic.

ROBERT: I'm not into sort of sweat-box, nutter-tops-off sort of stuff – that's just obscene! It's like, you don't want to be a fucked-up nutter in a room full of sad people! [we laugh]...then you come out looking like a schmuck. I mean, you want to be around...when you're really fucked then it's almost as if you want everyone else to be fucked too 'cos you are, but they're not [ummm]. You want everyone to be on the same sort of vibe. You're grooving to the music or whatever, kind of like more heads down and just sort of totally in your own world, your own...sort of head, just totally off on one sort of thing, which is cool sometimes, I mean, and other times you'll be totally more head up, totally looking at where the DJ is and the people will be jumping around you. If everyone else is on the same buzz and fucked or whatever [um-huh] and jumping about and having a laugh then that's going to be a good night...crowds DO make nights.

For Robert, the surrounding crowd is evidently crucial in the success, or not, of his night. In particular, there appear to be times and spaces when being 'cool' is about demonstrating control over one's body, while at other times being 'cool' is about abandoning or relinquishing control of one's body to the crowd. From Robert's story, it is apparent that for him these different times during the clubbing experience are again related to the tension between the ways in which a clubber can experience the crowd ('crowds DO make nights'). Both modes of experiencing the crowd – as distinctive from it (individualised) and as submerged within it (anonymised) – are able to foster moments of oceanic contemplation, but the nature of the crowd around him is implicated in influencing which of these modes Robert experiences. If Robert wants to interact with the crowd, it appears important that the crowd are 'with him', that they seem on the same level as him, that they are 'on the same buzz' and display broadly similar body techniques. On the other hand, if Robert is more reflective, more 'in [his] own world', then his head is 'down' and he is not interacting with others. Yet, this can be 'cool' too. What feels 'cool' at any one moment is thus at least partially crowd premised in an on-going and permanently unfolding fashion.

Although dancers may feel momentarily out of control of themselves, they may also experience this as a sensation of being partially controlled from *without*, by the music and the aura of the dancing crowd around them (Csikszentmihalyi, 1975b)[46]. This simultaneous sense of losing and gaining control over the dancing body is complex, particularly when related to clubbers' understandings of their own techniques and competencies. As Ludwig suggests, '[r]elinquishing conscious control may arouse feelings of impotency and helplessness, or, paradoxically, may represent the gaining of greater control and power through the loss of control' (1969: 14).

It would therefore be wrong to surmise, for example from Kim's description, p. 112, of her 'switching off', that the dancing crowd consists simply of a large number of dancers each attempting to 'lose it', to slip into automatic in their quest for pleasurable sensations. Apart from anything else, if that were the case then the dance floor would become a chaotic and dangerous situation resulting in injury. The dancers are not in a trance. Rather, the dancing crowd 'works together' in creating conditions for the positive experiencing of intensity and withdrawal, the oceanic, and, if drugs are involved, the ecstatic experiences. The body techniques of the dancers are orientated according to the changing demands of the dance floor and the social situation of the dancing crowd, as well as to their own senses of expression and release. The dancers are keen to avoid 'over-intrusive perceptual relations' with each other – hence, in Goffmanian terms, the feigning of 'disattention' and the occasional tactical use of 'involvement shields' to effectively go 'off-stage' temporarily. Yet the dancers must also – even in successfully appearing 'disattended' and experiencing the most intense moments of oceanic reflection – 'scan the scene of their action for postural, gestural and linguistic clues regarding the possible action or inaction of others, and they must provide similar clues themselves' (Crossley, 1995: 138). All this 'is necessary if the distinct projects and actions of embodied agents are to be coordinated...within shared spaces' (Crossley, 1995: 139). Differing areas of the dance floor demand different techniques of the oceanic just as differing times do, and thus the stereotypical 'abandon' of the dance floor is actually constituted through extremely complex and constantly changing socio-spatial negotiations and orderings.

Awareness of self is thus both heightened and in some way muted in the oceanic experience. Some control of the body is ceded to the music and the crowd, yet through this semi- or unconscious ceding a heightened sense of control over the body can be established. The apotheosis of this phenomenon is that clubbers can lose a sense of themselves as entities separate from the crowd – they feel lost in the crowd – yet they can use this sense of merging and harmony with the crowd as a context for interpretation and contemplation of self (Csikszentmihalyi, 1975c). Dancing clubbers are thus at times almost subconsciously attentive to the changing nature of the crowd around them, very subtly monitoring others' dancing techniques and territorial spacings.

JOHN: Uh...the black room, the big black room downstairs, I didn't find that until about one o'clock in the morning and I was totally fucking shocked when I went down there, it was mental. It was totally dark and there were just a few lasers here and there – red and black looked really good together, ummmm...and you didn't really get an impression of how big it was except when the lasers flashed, and then you sort of only got an impression and you couldn't really tell and that sort of gave it the impression of being a lot bigger than it really was...a couple of times I've done what I call 'lost it' while dancing, just totally [yeah]...but then I realise like five or ten minutes later, ummm...that you're dancing really hard and you see everyone around you dancing the same way – really hard, and actually I notice this girl looking at me like this and laughing, ummm...but that is the best part to me, of dancing, is when the music just builds up so much that you just go without thinking about it really, and...lose it. To do that it's got to be really really fast [ummm]. It just happens, you don't even notice it while it's happening and then afterwards you think 'fuck that was really good!' [yeah]. I was in the black room when Paul Oakenfold came on. It was...how do you describe the feelings...excitement, ummm, oh, contentedness [right] that, I dunno, that...do you know what I mean? [yeah]. I'd got something that I'd wanted almost and was just so nice – the heat, the lasers and blackness of it...that was awe-inspiring. It made you go huuuuuuuuuuuuuuuuuhhhhh [draws breath deeply inwards]. It was excellent.

John conveys here the oceanic sensation as he actually experienced it on the dance floor. John's experience of 'losing it' is partially based on the practices of the crowd around him, as well as his dancing and the building music to which he is responding. The fleeting nature of the experience is evident. Yet again, John's search for the words to evoke the feelings he experiences demonstrates the difficulty of articulating the nature of the oceanic.

In the following section of 'The night out', on 'playful vitality', I discuss in detail the exhilarating and often ineffable sensation of strength or vitality to which John alludes in this extract ('I'd got something that I'd wanted'). These sensations appear to result from this simultaneous losing and gaining of control over the body, through the feeling of merging with the crowd (or *exstasis*), and through the experiencing of 'glowing moments' (Laski, 1961: 65) or sensations of euphoria, spacelessness, timelessness and the momentary. Before this, however, I want to complete this section of 'The night out' by entering the chemically altered world of the ecstatic experience.

THE ECSTATIC EXPERIENCE

As well as being attained through the practices of dancing and the use of the body in crowd interactions and personal expression, the oceanic sensation of euphoria, liminality and *exstasis* can be experienced, prolonged or intensified through the use of recreational or 'dance' drugs. Where chemical triggers are used by the dancer, they are usually used in tandem with the practices of dancing and bodily control on the dance floor that I have just outlined, and thus the use of drugs represents in some ways an *additional* layer of emotional and sensational 'action', as opposed to an alternative. I want to turn now explicitly to examine the role of drugs, and especially the drug ecstasy (MDMA) in the clubbing experience, with a particular focus upon their use in attaining a form of oceanic experience that I am calling the ecstatic.

Ecstasy (MDMA) and clubbing

There is a whole gamut of drugs used by clubbers in Britain. Some are thought of as 'old favourites', such as cocaine and cannabis; others are newer 'synthetics', such as ecstasy (MDMA), MDMA-related substances, such as MDA and MDEA, and the anaesthetic-like Ketamine (often referred to as 'special K' or 'vitamin K')[47]. Still others are so-called 'legal highs' made from natural plant and herbal substances. Over the past ten to fifteen years clubbing in Britain has been increasingly pervaded by ecstasy (MDMA), and this has had a huge impact on the practices and nature of clubbing experiences. I am concentrating upon the use of ecstasy (MDMA) for two connected reasons. First, most of the clubbers I talked to who used drugs in clubbing currently used or had experience of using ecstasy (MDMA). Second, after cannabis it is the most used drug in Britain, and unlike cannabis is almost exclusively a 'dance drug'; that is, it is used predominantly during the practices of dancing during clubbing and raving (Saunders, 1997).

'Ecstasy' is the popular name for 3, 4-Methylenedioxy-N-Methamphetamine ('ecstasy' or 'E' is slightly easier in everyday conversation). It is a compound that was originally patented by the German E Merck Company in 1912. The popular perception that it was developed as a slimming pill is almost certainly a myth (Henry, 1992; Saunders, 1997). Saunders suggests that MDMA 'was just one of many compounds which were patented but never marketed' and which 'next came to light…in 1953 when the US army tested a number of drugs to see if they could be used as…agents in psychological warfare' (1997: 7). A Californian chemist named Alexander Shulgin re-synthesised MDMA in 1965, and after introducing a therapist friend to MDMA in 1977 the substance quickly became popular among West Coast USA psychotherapists because of its empathic effects (ISDD, 1996; Saunders, 1995; Shulgin and Shulgin, 1991). MDMA first appeared on 'the streets' of the West Coast of America in 1972 as, at that time, a legal alternative to MDA, which was a related substance that had already been made illegal. In Britain MDMA or 'E' was initially associated with clubs that played so-called 'Balearic' dance music in the mid-1980s, although its possession was made illegal much earlier than this through an amendment to the Misuse of Drugs Act (1971) designed to blanket-ban all amphetamine-like compounds (ISDD, 1996). MDMA is thus a Class A substance (Schedule 1 in the USA, where it was banned in 1985) – a designation reserved for those drugs deemed to be most harmful.

The use of MDMA can induce a feeling of euphoria and benevolence. However, while it is believed to enhance perception, its psychedelic potential is low, at least in its pharmacologically pure state (a rare occurrence – it is almost always blended with non-MDMA 'fillers'). The positive effects of boosted energy levels, confidence, happiness and heightened empathy that MDMA can provide can be offset by blurred vision, nausea and vomiting (Release, 1997; Stevens, 1993). In Britain the drug is taken orally as a tablet or capsule with an MDMA content of usually between 0–150mg, although higher amounts have been reported (Saunders, 1997). The street price for an 'E' (or 'pill') is currently (late 1998) between £10 and £15.

Given that our bodies express the emotional states in which we find ourselves (Frith, 1996) and that we are able to feel 'oneness' or 'sympathy' (Scheler, 1954) with others in certain specifically framed situations (Blacking, 1973), it follows that the use of a drug which particularly impacts upon emotional, and thus also physical, states can affect the sense of identification that one may feel (or not) with a clubbing crowd. However, clinical and psychological research on the effects of so-called 'dance' drugs is still scarce. What little work has been done suggests that 'empathogens' such as ecstasy (MDMA) undoubtedly affect an individual's sense of self – both their identities and identifications – with some studies suggesting that 90 per cent of ecstasy (MDMA) users experience a feeling of closeness to others (Release, 1997).

Recent studies have suggested that between 68–76 per cent of clubbers regularly take ecstasy (MDMA), with many others taking amphetamines (such as 'speed') and LSD or smoking cannabis or 'grass' (Mullan et al., 1997; Release, 1997; Solowij et al., 1992). For example, Mullan et al. (1997) surveyed 720 people at club events between April and September 1996 – overall 92 per cent of those surveyed had used drugs, with 76 per cent having used ecstasy (MDMA). Ecstasy (MDMA) seemed to be used more on a weekly rather than a daily basis (as cannabis was), with approximately half of those having taken ecstasy (MDMA) in the previous six months taking it every week. Of those who completed the questionnaire, two out of five were using ecstasy (MDMA) every week. In the Solowij (1992) survey, 68 of 100 people (68 per cent) surveyed had used ecstasy (MDMA) more than three times, while 83 (83 per cent) had tried it at least once. While the Release Survey (1997) found that 81 per cent of those in their survey (clubbers aged between 16–29 years old) had tried ecstasy (MDMA) and 91 per cent had tried cannabis, the British Crime Survey (BCS) (1994) indicated that, of the population of 16–29 year olds *as a whole*, only 6 per cent had ever tried ecstasy (MDMA) and 34 per cent had tried cannabis. This indicates the highly unusual levels of drug use that can occur at clubbing events. Interestingly, the British Crime Survey (1994) also noted that 16–29 year olds who went to pubs, clubs and wine bars were nearly twice as likely to have ever taken illegal drugs than those who did not go to these places, and three times more likely to have taken drugs in the last month – 'in essence, people who go out take more drugs than those who stay in' (Release, 1997: 12). A survey by the University of Exeter (1992) suggested that 4.25 per cent of 14 year olds (24,000 young people nationally) had tried ecstasy (MDMA) – a worrying statistic that reinforces the need for effective and co-ordinated drugs awareness education to be available from a young age.

The British Crime Survey (1994) suggested that as a whole between 303,000–385,000 people were regular users of amphetamines (a group of drugs that includes ecstasy [MDMA]), and that between 2,486,000–2,696,000 people had used amphetamines at least once. Overall, 1 per cent of those under 30 years old and less than 0.5 per cent of older people were found to be regular users of ecstasy (MDMA)[48]. So while many of those who go clubbing or raving have tried and continue to take ecstasy (MDMA), it should be stressed that among the general population by far the 'norm' remains not to have tried *any* form of illegal drug *ever* (Balding, 1997; cited in Druglink, 1997). The media-fuelled moral panic that surrounds drug use in Britain, especially use by young people, often obscures this basic fact[49].

A further point relating to the use of ecstasy (MDMA) in the clubbing experience is that within general media and clubbing discourses the term 'ecstasy' has become almost synonymous with any unknown street drug in pill form. According to one study, it could mean 'anything', with the label applied to virtually any drug (ISDD, 1996): 'what is sold as ecstasy in Britain may in fact be a number of different drugs' (Saunders, 1997: 212). While most pills have at least some MDMA in them (Saunders, 1995; 1997), others are constituted from, at best, MDMA-derivatives[50] and, at worst, cheap 'fillers'[51]. Thus, while clubbers refer to 'ecstasy', the pill or capsule that they have taken may have contained none or very little MDMA, with the chance of getting 'pure MDMA' in Britain being about two out of three and the chance of getting a pill containing a similar type of drug (such as MDA or MDEA) being about three out of four (Saunders, 1997).

The use of drugs in the attaining of states of ecstasy and as a fast-track into the experience of the oceanic is far from new. In fact, it can be traced back over thousands of years (Abel, 1980)[52]. There has been a long association in Britain between all-night dancing and the use of drugs, going back to the use of stimulants such as cocaine in the pre-war West End club scene (Cohn, 1992; ISDD, 1996; Saunders, 1997) and including, for example, the 'furious consumption' (Barnes, 1980) of amphetamines by mods during the late 1950s and 1960s. The scale of the use of recreational drugs in contemporary clubbing cultures is, however, unprecedented.

A NIGHT ON E: THE USE OF ECSTASY (MDMA) IN THE CLUBBING EXPERIENCE

One way in which to represent and attempt to understand the role of ecstasy (MDMA) and other dance drugs in the clubbing experience is to 'shadow' clubbers who use drugs over the course of the night out. Like the practices, spacings and timings of the night out more generally, the practices of taking drugs – whether ecstasy (MDMA) or any other drug – are complex. Skills, techniques and notions of competency and coolness in the practices of finding, buying, preparing for, taking and coping with the consumption of the drugs are important in the constitution of these ecstatic experiences.

Clubbers who do take drugs as a route into the oceanic experience will each have their own personal routines and rituals which will vary from night to night. However, there are specific practices and sensations that are commonly expressed, with the spacings and timings of drug consuming practices of the dancers taking on a relatively ritualised form. The brief schematic that follows is therefore merely an evocation of a night on E based on nights out with clubbers during the course of the research for this book. The complex and highly ritualised nature of the consumption of ecstasy (MDMA) during clubbing merits half a dozen books all of its own, and thus what follows can, in the space available, only drift lightly, yet hopefully suggestively, through the ecstatic experience.

Pre-clubbing – sorting, preparing, bonding

The period immediately prior to clubbing is always exciting – drugs or no drugs. For those intending to take drugs in the hours that follow, this excitement is often augmented by a sense of nervous anticipation and the intricacies of deciding what they want to take, where and whether they are going to get it, as well as by the less-discussed worries that they may have about their forthcoming experience: Is the drug 'safe'? Is it 'E'? Will I be okay? Will I get it through the security at the door? Will my friends be alright? These latter worries will be magnified for those that haven't experienced using ecstasy (MDMA) before.

Concerns about safety are reflected in the efforts that some clubbers make to research what they will be taking. For example, the clubbing magazine *Eternity* features reviews and news of different forms of ecstasy (MDMA) tablets, giving information on the results of tests. Nicholas Saunders' Internet web-site also features up-to-the-minute news on ecstasy (MDMA) tests and 'dangerous' pills with detailed breakdowns of the chemical constitution of current types of E. This web-site was consulted up to 400 times a day during 1996 (Saunders, 1996). For the extremely patient, the Drug Detection lab in Sacramento, California, does postal testing of pill samples for $100 (Saunders, 1997), although the turn-around time and cost means that use of this service is extremely unlikely for British users.

VALERIE: I'm not going to do loads because then I would worry, yeah, but if I did a whole pill to begin with I know I would be paranoid, so there's no way I'd fucking do it in the first place. I'd be worried about how fucked I'd be. I think I'm a lightweight. It does make me laugh sometimes though, 'cos I think to myself there's like people taking three pills and here's me with my half. I think I did worry a bit about the effects on my brain when I first started taking it; long-term effects. I was concerned. When we started taking them we looked up on the Internet every bit of information on Es to find out what, roughly, the effects were [uh-huh]. We tried to work out what a safe dose was, we tried to work out what we were taking, yeah [uhmmm] 'cos there's so many different sorts. We tried to get as clued up as we could on it...I really insisted on it. I think everyone else was like pretty like interested, but I was definitely wanting to know what I was taking. I've realised through the last year that it's impossible to know – you just can't know. In a

120

way, I know it sounds strange, but I get ill on them, I know it sounds a bit peculiar...Okay, this is the story: I take E twice in a row on weekends and I get ill, and I take E sort of every month or a little bit longer and I'm okay, so from that I've deduced that I have to be really careful with my intake.

BEN: Do you not worry about taking it at all, in that case?

VALERIE: I might do – if I have kids then obviously I won't do it while I'm pregnant, and I won't do it while I'm breast-feeding so that's like two years of not doing it...you know they've got those testing points in Amsterdam – they've got like 10,000 bits of data. Do you know what I mean? You could be getting anything, absolutely anything [um-huh]. I mean I think some people have all sorts of problems – liver problems, whatever, and they just can't handle it. There's an amount of people that get admitted to hospital every week, not through E-ing but just through overheating.

From this quite personal exchange with Valerie, it is obvious that she clearly spends a great deal of time and effort worrying about exactly what it is that she is taking when she goes clubbing. Despite Valerie's apparently carefully regulated intake of the drug ecstasy, it is evident that, like the overwhelming majority of those in her position, she really does have no idea about what it is she is taking each time and what the long-term effects on her might be (and, realistically, how could she have?). The subsequent importance of the group-based decision-making process is also clear. For most clubbers, decisions about whether to take drugs, where to get them, how much to take and when to take them are indeed made as a group. This form of decision making appears popular because it gives the illusion of taking the responsibility for choices about taking drugs away from the individual clubbers. Of course, ultimately these choices remain individual ones. This early group involvement engenders sensations of 'not being alone' in both the anticipated risks and the excitements that await during the night ahead, and thus before the night has even begun, notions of group trust and bonding may be developing, particularly if (as is usual) previously successful nights are being evoked and used as a foundation, framework or rationale for the night to come.

BEN: So it's normally a relaxing afternoon?

KIM: Yeah, maybe pop into town, go out and get something to eat, again it's really just sitting about, talking, reading, having a relaxing day, but not sitting about smoking [cannabis] too much or otherwise you're just dead by the evening...

VALERIE: ...totally.

KIM: ...so we try and do something like just go into Dartford and do a bit a shopping, cook dinner and that sort of thing, ummmm...sort of early evening the music gets turned up and people will be popping in and someone will be sorting out the pills and stuff. And it all starts getting a bit technical...

VALERIE: ...when we sit there cutting up our pills, that's one of my favourite moments. We all sit there and we've all got our cling film or whatever and we're sitting there going 'How are you going to do yours? Halves or quarters? What time are you taking them?'...we sit there and have this massive discussion.

KIM: We used to do that a lot more when we first started going out. It would be...we'd all do it at the same time, but now it's more of an individual thing...

VALERIE: ...I think we've worked out that we've all got a different pace, yeah? [right]...and before we were all doing it together, but now it's much more individual. I think it's partly because we all feel a little safer with it as well. We used to get all our pills beforehand, before we went up there, but now, we're having trouble...I mean our theory was that we weren't going to get any shit off people that we knew, yeah? [right] We're not going to get any dodgy pills.

A glimpse into the pre-clubbing routines of Valerie and Kim is provided in this extract. The early evening rituals of relaxation are gradually replaced by a set of increasingly 'technical' decisions and operations concerning the sourcing and purchasing of the drugs, the amount each of their group will be taking and the implicit worry that the clubbers have about the actual content of the drugs. Despite Kim's claim that 'it's more of an individual thing now', the social nature of these decisions and of this stage of the early evening as a whole is clear.

As far as getting hold of the drugs is concerned, the most common source is a friend who 'does not sell regularly' (in other words, is not a 'dealer'), although 'almost as prevalent [is] buying from a dealer' (Release, 1997: 15). According to the Release survey (1997), buying from a friend became even more popular with increasing age, with those aged over 30 years nearly twice as likely to have bought from a non-dealing friend as opposed to an unknown dealer (ibid.: 15). This progression highlights the frequent exposure of especially younger clubbers to unscrupulous and/or unknown and usually older people.

'Dropping' the 'pills' – taking the drugs

It is probably safe to say that no two clubbers have the same method or go through the same mental processes in taking ecstasy (MDMA). While clubbers might take the same amount and even the same 'brand' of pill, each will approach the experience differently. Some may 'drop' part of what they intend to take even before the night has started – maybe at home or more commonly while waiting in the queue outside, particularly if it is already late or they are very near the front of the queue. However, it is most usual to 'drop' the first amount, of what might be just one but may also be two, three or even more amounts over the course of the night, within a short space of time after entering the club. It usually takes between twenty minutes and an hour to start experiencing the effects of the drug – to 'come up' on it. Conveniently, this leaves time for the clubbers to find a space and to have a drink, deposit coats and bags in the cloakroom, and wander through the club generally getting into a clubbing mood, checking out the music and perhaps meeting friends. In instances where clubbers buy the drugs ('score' them) when they get into the club, thus avoiding the risk of bringing them through the security themselves and the potential difficulties and dangers of scoring the drugs beforehand, the search for a dealer from which to 'score' may also take place in these initial stages. Occasionally, perhaps due to a later arrival, a build-up of excitement or a familiarity with the club night, clubbers will 'drop' their pills immediately on entry and go straight to the dance floor, with the nervous excitement of having just taken their pill combining with the music and their familiarity to induce an impatience to start dancing and for the night to begin.

BEN: Everyone was going crazy already…at half eleven! What was going on?
VALERIE: I tell you what, when I got in there, something…I dropped my pill because I was really eager to get my pill down me so I could come up straight away yeah…and ummm…when we got in there I started dancing straight away – it was just like I was bopping away from the moment that I got in there and I don't usually do that. I like to sit down usually. I was SO excited. I was like a fucking kid in a sweet shop!!
BEN: I was a little surprised at how quickly everyone started going for it…
VALERIE: I was really eager to just get on with it, I just wanted to get 'up there'.

This impatience that Valerie exhibits in taking the drug is common. The pre-clubbing period of playing loud music at home or in the car, perhaps smoking cannabis, and generally getting wound up for the evening inevitably leads towards two key moments in the early part of the evening. The first, and one that characterises nearly all clubbing experiences, whether or not drugs are involved, is the initial, overwhelming and nearly always adrenaline-producing experience of the music at the club. Much louder and more powerful than anything that might be produced on a home- or car-based system, this music is as much felt by the whole body as simply heard. The second key moment is restricted to those that take drugs such as 'E', and that is the moment of actually taking the drug. This is a vital moment that results in a build-up of tension, partly because of uncertainty over what is actually in the pill the clubber is taking and partly due to anticipation of the excitement, euphoria and adventure to come.

There is no 'usual' amount that is taken, and, in one sense, it matters less how many individual pills are taken and more what the effects are. Sometimes a half or a single pill will be sufficient for the clubber to experience the sensation that they are seeking, whether or not this is an ecstatic experience. At other times clubbers may take three or four pills. One clubber with whom I went out took four within a single hour, complaining of 'poor quality'. Furthermore, the size and weight of the clubber's body and the vital issue of tolerance will influence the amount of the drug that is needed in much the same way that a unit of alcohol will affect drinkers differently.

First experience as revelatory

Clubbers often talk of their first experience of ecstasy (MDMA) and other 'dance' drugs in revelatory terms – almost as if they *only now* know the score or 'the secret'. It is usual for clubbers to talk about their first time with a strongly positive tone, even for those who have long since ceased to take the drug.

SEB: I've experienced the friendliness of clubs when I danced in Glasgow, but because I wasn't into ecstasy I didn't appreciate that that was what it was. I just thought that the guys liked me, I was that naive – guys would shake my hand and I thought, 'brilliant!'. I was totally into it, I loved the music and I loved the dancing and I knew I was enjoying myself so much that I just thought they were as well, and if anything I thought that they were enjoying it more than me so that just egged me on to enjoy it harder. I didn't realise that when you drop a pill you just start smiling and it all takes over.

Seb presents a picture of *pre*-ecstasy and *post*-ecstasy; the former characterised by what he sees as his naivety, and the latter by the revelation that all was not as it seemed, that now he 'knows'.

'Coming up' – starting to lose touch

Usually, even for relatively experienced users, once the pill has been taken a short period of nervous waiting follows during which the clubber will discover firstly whether or not the pill is a 'dud' – there is no effect or unusual and unpleasant effects – and secondly, if it proves not to be a 'dud', the strength of the pill. Feelings of nausea are not uncommon, in part due to nerves, in part to the physiological effects of the pill dissolving, and the clubber may experience a feeling not unlike 'butterflies in the stomach'. Occasionally clubbers may vomit, yet these feelings of sickness usually pass within thirty minutes and, in most cases, within an hour the clubber begins to experience an altered state of sensory perception, a surge of euphoria and elation, feelings of unbounded energy and a heightened sense of empathy with those in the crowds.

BEN: Okay, so where were we – you were telling me what E is like for you...

JOHN: Okay, um...after about twenty minutes you start to feel the edges where you start to feel something, feel slightly different, or I do anyway [um-huh]. Usually smoke a joint around then – I try to if we're at a club, hopefully we've got a joint with us – we head off to the middle of the dance floor if there's nowhere quiet – smoke it there and that brings us up. What do you notice ummm...uhhh...altered perception is what I really notice, it's the fact that I'm noticing things differently...something has happened ummm...and I start feeling very excited as well, and like light in my stomach, and light in my head as well then, and I dunno...it's like, here we go, it's just all starting, and that's when I start listening to the music as well and that builds up, and not this time but the time before at Banana Split, I got in there, had a pill, had a joint and then went up to the dance floor [right] and we were standing about just looking at each other as the first tune was building up and the main thing that was building up was the atmosphere between us as the tunes built up, it was very trancey sort of stuff and just everyone was looking around grinning at one another and thinking right here we go, this is going to be a quality night [right]. They know where you are and you know where you are and you both go to the same place. While I'm on the pill I feel absolutely fine, I find it very easy not to worry about anything – I've never had a paranoia attack although a few times I have been a little bit ill – felt sick, been sick and then afterwards I've just been saying to myself 'no, I'm fine, it's just me winding myself up, I got too hot or whatever, I hadn't drunk enough' [yeah, yeah]...ummm...one of the things when you're on E, you can't control your body temperature and you start to feel a little bit sick, and then if you don't immediately go into a chill-out area and try and cool down a bit...for me anyway...I start feeling it and I start thinking I'm going to throw up and once I start feeling I'm going to throw up that's it, I'm going to throw up. I think that's more attitude than the actual drug itself – it's more me thinking 'Oh I'm going to be sick, I'm going to be sick'...I mean a lot of the times I go running off to the toilet, into the toilet and it's cool in there and I feel fine [yeah], or even when you do actually heave you only heave once and it's like errrrhghghghh! [making 'sick' noises]

The build-up of tension that the clubbers share in the early pre-climactic period of the night is evident in John's story. A feeling of familiarity and of sharing ('here we go again') and the sense that they will shortly be 'going' somewhere, to another 'place', plane or level, are characteristic sensations, particularly among groups of friends. The clubbers feed off the emotions of each other, developing 'the atmosphere between us'. An ethos of sharing something extraordinary evolves between those within the dancing crowd, and this provides the starting point for an ecstatic sensation.

Stripping of defences – a 'natural' state?

'It made me feel how all of us would like to think we are anyway'...ecstasy can act as a reminder of a kind of honesty rarely found in human relations. Certainly, nobody needs a drug to tell them of this, and it may be that the message that Adam brings is the reminder that ecstasy has always existed without it.[53]

(Nasmyth, 1985: 78)

A recurring feature of the experience of ecstasy (MDMA), as well as other *non-drug* induced oceanic experiences, is the sensation that one has witnessed the revelation of the 'true' nature of one's self and of others – the individual 'suddenly' sees how people 'really' are, their 'natural' states. Stevens describes this sensation as a 'melting [of] defenses [sic]' (1993: 488). It appears that the ecstatic experience facilitates what can seem like a fleeting glimpse of sanity or 'naturalness', as the inhibitions of an apparently rule-bound 'outside world' dissolve in the space of the dance floor and within the dancing crowd.

BEN: How do you feel about yourself when you're there on the dance floor?

VALERIE: I believe that we all have personas, yeah [uh-huh, right]. So you're a totally different person with me than you are with your mother [oh, definitely!]. Right, and...I think I like to believe that when I'm on an E I have no defences whatsoever, so in many ways I would say that the person I am when I'm on an E is the 'real me' right, because I feel totally open and...I feel clean, yeah...I feel cleansed of all my worldly woes.

BEN: What do you mean that you have 'no defences'?

VALERIE: I don't...I don't worry yeah...I don't, I feel like, I trust, I feel more trusting, I'm quite a trusting person anyway, yeah. I don't...it's really difficult – I feel really sort of like spiritual like I don't judge, I trust, I feel cleansed, I feel...it's like a really sort of pure feeling.

BEN: But is that different from others parts of your personality?

VALERIE: When you're at a rave and you're with hundreds of people dancing and you're having fun yeah, you don't worry about what your mother said to piss you off earlier that day, yeah? You're just concerned with what's going on then – there's a really happy atmosphere, a really happy vibe, and you're part of that, yeah. When you're not in that setting it's not that you've got to worry about things, but you've got to…not be on your guard but you've got to be aware of things yeah…there…there are people who for their own reasons are doing what they do, yeah, which are right to them, but they don't necessarily tie in to what you want, so you've got to be aware of it so that you can compromise, so that you can analyse, so that you can do whatever, yeah. When you're on an E, everyone's there to enjoy themselves.

BEN: So you feel that you're amongst like-minded people?

VALERIE: The thing is I think it's easier because you know that everyone else is also on an E, yeah, so you're getting the same sort of signals sent out, so…

BEN: How can you tell if others are on a drug?

VALERIE: God, I don't know? I really don't know. That's quite strange really because I don't know what people are on. Like Brian, he said people were on coke [cocaine] there, yeah…at Banana Split, and I hadn't noticed. He said that people going like that [makes face]…and I hadn't thought about it, I just assumed that everyone was really happy, so it's sort of like my perception of it more than anything else [yeah]. I perceive everyone to be happy. I mean if I smile at someone and they don't smile back I just think they're really fucked and I think enjoy it for what you're at, at that moment [yeah].

The non-confrontational, consciously open attitude that many clubbers suggest forms one of the most attractive aspects of clubbing, is discernible in Valerie's clearly carefully articulated account of the 'different' persona that she presents on the dance floor. This sensation of being able to identify a 'true' self, of finding who you 'really' are, is one that Valerie appears to value highly. 'Defences' that Valerie deploys in other social encounters appear unnecessary while dancing, for having taken ecstasy (MDMA) Valerie has no worries but rather, 'trust, I feel more trusting […] I feel cleansed'. It appears important to Valerie that others within the clubbing crowd are also there for these reasons – 'everyone's there to enjoy themselves […] you're getting the same sort of signals'. The notion of 'momentary time' and its relationships with identities and identifications is also further textured through Valerie's description: 'you're just concerned about what's going on then […] enjoy it for what you're at, at that moment'.

Far from being concerned with some form of mindless and meaningless hedonism, then, as often portrayed in popular (mis)representations of clubbing[54], it seems that the experiencing of ecstatic sensations can actually be about an extraordinary and, for many, unparalleled and extremely precious experience of *their own identity*. Crucially, this experience of identity is perceived as their 'real' identity – how they really are (and/or want to be). This dramatic sensation of having one's 'true' self or 'natural' state uncovered, unveiled or released can be so

127

powerful as to be experienced as a form of earthly utopia or dream-world. This is a notion that I broaden in discussing the notion of 'playful vitality' in the next section of 'The night out'.

Elation and euphoria – experiencing the ecstatic

At their peak, ecstatic experiences can culminate in the temporary loss of a sense of time and space, and the experiencing of a strong sense of alterity and euphoria, as the dancing clubbers experience the tension of being caught between the dancing crowd and their self, between an atomistic sense of identity and a sense of (crowd) identification, between the urge for outward expression and the opportunity for inward reflection, between the music as controlling them and themselves as in control, between isolation and community. By no means is this ecstatic experience automatic – drug use does not guarantee an ecstatic experience, but can make one easier to experience, and longer in duration and greater in intensity if it is experienced. Unpleasant side-effects of the consumption of ecstasy (MDMA) are not uncommon, although for most clubbers the positive effects appear to outweigh easily the negative effects (Release, 1997).

BEN: Perhaps a tricky one – what does ecstasy actually do for you?

BRUCE: That's a very tricky one, ummm…it makes me feel utterly elated, uhmmm…it makes me very relaxed and at the same time incredibly, you know…I'll have bags of energy which is an unusual feeling to feel, uhhmm…and it, I don't know, I certainly don't hold any magical qualities to it, you know, it's familiar enough now for me to know what to expect [right] but I feel…

BEN: Do you treat others differently at all?

BRUCE: Yeah, you are far less judgmental and you're certainly far more tolerant uhmmm…and you're far more passionate and you're prone to saying 'I love you' uhhmmm and…but you know it just makes you feel good, you feel very…if you're going to something like Che Guevara where I know then it makes you feel very safe as well. You don't feel like you're going to get busted at any second, you don't feel like something bad's going to happen because you know how to handle this feeling and it's a feeling that you have actively encouraged.

This tension between energy and relaxation that Bruce describes is an attribute of oceanic experiences – both drug- and non-drug induced – more broadly (see Laski, 1961; 1980). Yet, this short extract is characterised by the notion that Bruce is in control of himself and his emotions – he is relaxed, elated, safe, tolerant – he knows how to handle the feeling. It is no coincidence that Bruce is aged 26 while John (in the preceding extract, p.125) is aged 22. The extra years of experience, while not guaranteeing control, are likely to mean that the older clubber has more refined practices of drug-taking (and clubbing), knows his or her limits and, in one sense, knows better what to expect as well as 'how to behave'. This routine nature of clubbing appears vital to Bruce's enjoyment.

Intensity and withdrawal

If s/he does experience the ecstatic sensation, the clubber may experience it *intensively*, participating within the crowd, actively interacting with others and experiencing euphoria socially. On the other hand, s/he may experience the ecstatic through *withdrawal* into themselves – through reflection, interpretation and sensations of clarity and tranquillity (Laski, 1961). This continual fluctuation is palpable in the extract where Dionne is describing the sensation of 'coming up' on a mixture of ecstasy (MDMA) and speed (or amphetamine).

DIONNE (AN EXTRACT FROM *The Moment*[55]): Wander around for a bit, chat for a bit, can see a few familiar faces so that's cool. Seems a good crowd and not so young – paranoid about being over twenty-five. Think I'll just stand still here for a minute and lean my head against this pillar, close my eyes and let the sounds wash over me. The bass is pumping, I can feel it reverberate in my body...so that's where my sternum is. I like this. Time to dance I think, just casual to start with, check out the opportunities. Couple of raised-up plat-forms with good showing off potential for later on. Ditch the beer and fill the bottle with water from the cistern, can't seem to get the taps to turn on. Not a smoker but now's the time for cigarettes. Gum too, starting to gurn. No pockets so have to pester boyfriend every time. He gets a bit fucked off and that brings me down for a while. Shit, don't need this, what I need is another pill and then I'll just be ready and waiting for the moment. Do a whole one, feel a bit guilty for quarter of a second then head to chill-out room. Find Sarah – whispers conspiratorially how much have you done, 2, that's alright then as she pops another. Rushing big time now and sit down heavily leaning against the wall. Mouth runs away with me as I insist I've seen this woman on TV – get a grip. Steve comes and sits down, he's well loved-up. He puts his arm around me and I grip his arm tight. It's suddenly very important I must hold on. Don't ever let go. He smiles. My heart is racing and I scrounge a fag. It – the moment – will be soon.

Whether the clubber experiences intensity or withdrawal at any one moment depends upon a number of factors. To give a simple example, if it is dark and crowded and visual contact with others is problematic, withdrawal is more likely than if there is more light and eye contact can be easily made. Moreover, a fuller dance floor may trigger an experience of intensity with the crushing tactility of the crowd-inducing euphoria and a sense of closeness, yet it may equally trigger withdrawal through the employment of an 'involvement shield' (Goffman, 1963: 38) as the dancer gets 'lost' in the crowd.

Second, the clubbers will orientate themselves within the regionalisations and mediations of the dance floor depending upon the spacings and timings of that stage of the night, the physical configuration of the club, especially the location of the DJ booth, their own location and whether they are within a sub-group or not, and the type of music being played at that moment. If the DJ is visible, perhaps on a raised stage or platform at the 'front' of the dance floor, then the clubbers will be more likely to face him (or, less usually, her). If the dance floor is relatively small or the periphery too well lit, then the clubbers will be more inclined to face *into* the dancing crowd in an attempt to avoid eye contact with non-dancers, who, it might be assumed with some confidence, are not on the same motional and emotional 'wavelength' as the dancers are at that moment. If dancing within a group of friends, the clubbers may be more likely to face other members of that group. If alone, they will be more likely to face the 'front' of the club, perhaps where the main lighting rig or slide projector images are positioned. These are further examples of the 'engagement closure' that Goffman mentions in his discussion of crowds and social interaction (1963: 156). In any case, any environmental feature – physical or social – which encourages crowd interaction on the dance floor will be more likely to foster an intensive interaction as a result of the ecstatic, with the reverse (withdrawal) being the case in instances where crowd interaction is difficult.

Of course, a third factor that impacts upon the clubbers' experience of the ecstatic is their pre-clubbing state of mind and attitude – the impact of their social lives more broadly. Unquestionably, in certain cases where I accompanied clubbers on their nights out, my presence had a tangible impact upon their state of mind and thus their enjoyment of the night. However, it is difficult to qualify this any further without making assumptions about the clubbers' practices when I was not there, which I am not prepared to do.

The music is a fourth feature of the night that impacts upon the nature of clubbers' experiences of the ecstatic. The playing of 'special' musical tracks – what are sometimes called 'anthems' or 'classics' – can trigger a very sudden shift across the dance floor from predominantly withdrawal to mostly intensity. An introverted experience can quickly become an extroverted one. Catching a simple glance or smile from another dancer while dancing can also suddenly pull someone quickly away from a sensation of personal, inwardly orientated reflection and isolation and into a crowd-based sensation of empathy, intensive inter-clubber interaction and a sharing of spaces and emotions.

BEN: So, as far as I understand it then, for you, you take ecstasy for stamina, but also for other reasons...?

KIM AND VALERIE: [together] Oh yeah...

VALERIE: Someone will say something to you and you'll go: 'Yes, definitely' – you're totally connected.

KIM: And also, you don't need to talk. Part of the thing of going out clubbing is that you've got everyone around you...this is what I was saying about expressions – everybody is expressing themselves. Take Tim and that – you [talking to Valerie] look over and whereas I haven't seen you for sometime and we may have got a little different, when we're at a club we might look over at each other and we both grin because we don't NEED to say anything. It'll be that connection. It's all there. Partly, it's experiencing things together, whatever they are, but because it's on such a different level and everything's so new and...umm...

BEN: ...and do those feelings extend to people who aren't in the immediate group, beyond your friends I mean?

KIM: You might not know what they're thinking, but you know that they're there to be happy...

VALERIE: ...but I definitely feel a lot more empathy for our little group than...when I see people in a club I just think excellent, they look really happy, they know exactly where I'm coming from, yeah, but when I look at Rachel, David and Tim, there's something more...

KIM: ...but then you can look at the other people in the club and it's more than just them being happy. It's more than that. You're dancing to something and it's building up and up, and you look around and they're building with it as well, when...it's the experience you have together that matters. You do sort of empathise and connect better with them. I think it's really natural.

VALERIE: When I'm straight it's a different experience – I don't talk to people when I'm straight. I'll just go to dance, yeah? [right]...and I'll look at people and I'll know how good they're feeling, yeah, but I'm not on the same wavelength as them...

KIM: ...you don't feel as in tune. When you're up there dancing and you're on one you can get going as the music will get you into it and you'll be really going for it, but when you go and sit down and like take a break and just sit down and look around, sometimes you'll feel part of it, sometimes you won't though as you're more an observer of what's going on than part of it [yeah].

This wonderfully nuanced and evocative passage is interesting for many reasons. I want to make four points. First, the clear sense of empathy and of a shared ethos existing between clubbers ('you're totally connected') is obvious. This is manifested in the lack of the need to say how they feel – they just 'know'.

Second, the importance of the group of clubbing friends is clear; while Kim and Valerie feel empathy for others ('you might not know what they're thinking, but you know they're there to be happy'), what they feel for their close friends is 'something more' than this. Third, the 'natural' quality of the empathy that Kim suggests the non-verbal communication of the dance floor can generate is suggestive of the notion of a 'real self' which, as I proposed earlier, appears so valued in the clubbing experience. This 'natural' feeling is generated through being within the dancing crowd – 'it's the experience you have together that matters'. Finally, Valerie notes how her dance floor experiences of being straight are quite different from her experiences of being on 'E' – she is not on the 'same wavelength' as others, but just goes to dance. Kim adds that she feels more of an 'observer' when she is not on 'E' – 'you don't feel as in tune'.

What goes up...

Just as the ecstatic tide comes in, so it must recede, and inevitably the effects of the ecstasy (MDMA) will begin to wane. The post-ecstatic experience shares many features with the post-oceanic, although the 'come down' often lasts longer and can be far less pleasurable than during the post-oceanic experience. Furthermore, a post-ecstatic experience may be accompanied by physiological effects such as tiredness, edginess, stomach cramps and insomnia. This is in addition to the usually positive mental and emotional effects that characterise the 'afterglow' of the oceanic (Laski, 1961). I discuss this 'afterglow' further in the final part of the book, 'Reflections'.

THE ECSTATIC AND THE OCEANIC IN CLUBBING

Whether reached partially through the use of drugs or not, during an oceanic experience the (usually) dancing clubber will perceive music, themselves and others within the dancing crowd in an altered, non-everyday fashion. Even when not fully attained, the mere experience of oceanic sensations in *previous* clubbing nights out appears to impact upon the attitudes, practices and emotions of the dancing clubbers at that moment. Clubbers' understandings of themselves and of their own identities are transformed during these moments of ecstasy. This transformation may endure long after the experience itself.

The role of the music in the oceanic experience should not be overlooked in the quite justifiable focus upon drug use. The use of drugs such as ecstasy (MDMA) can change the way that music is appreciated (Storr, 1992). It is also clear that a dancer's appreciation of rhythm and bass in particular can be heightened through the use of certain drugs, almost to the extent of inducing a form of trance (Inglis, 1989; Rouget, 1985). Yet, at the same time, there is no evidence to suggest that this trance experience is not also attainable *without* the use of drugs (Laski, 1980). In this study it is only possible to say with any confidence that it certainly appears feasible to experience music in a heightened form and as inducing an oceanic experience through dancing *without* the use of drugs at all. Finally, it is clear that clubbers' understandings of themselves and of both the wider dancing crowd and those who constitute it are altered during, and as a result of, the oceanic experience, as well as through ecstatic experiences which involve the use of drugs.

So far, I have only briefly touched upon the role of notions such as 'play', 'escape' and 'utopia' with which clubbing appears to be entangled. These notions have been bubbling just below the surface throughout this section of 'The night out'. In broadening understandings of the attractions of and roles for oceanic and ecstatic experiences in the everyday lives of the clubbers, as well as continuing to elaborate upon the spacings, timings and practices of clubbing more narrowly, I turn now, in the next and final section of 'The night out', to the notion of 'playful vitality'.

VALERIE: Everything seems okay, everything seems fine, I…quite often when I'm raving I think to myself 'I'm so lucky', yeah, I just think, 'I'm the luckiest person in the world to have the opportunity to do this, to have the friends I have around me, to have the people I have around me, to have the club I've got around me' and I just feel like the luckiest person in the world, and I sort of want to cry, yeah, I want to cry with happiness. I never sort of physically want to cry, but you know that feeling [yeah] – total joy yeah, oh, it's like…ecstasy! It's total ecstasy!

You haven't got the everyday, well, if you call it pressure, I mean some people are a bit pressurized more than others. I'm lucky. I'm not too pressurized. But at least you can come here and forget the whole damn lot. It's right out of your mind. You've forgotten work. You've forgotten even the street you live in. You've forgotten your next-door neighbours. Just for a little while. You can think about nothing, if that's possible. [56]

We're going to places
that are new, and old,
and blue and gold,
and rainbow bridges carry us
to another side of dreams,
we're swiftly moving galaxies away;
what a place to play,
what a place to play.

(Marlena Shaw, 1977, 'Look at me, look at you')

CLUBBING AND PLAYFUL
VITALITY

There is a gregarious impulse, as it were, that pushes one to seek out others, to touch them, and incites one to get lost in the mass as if in a more vast entity where one can express, by contagion, that which enclosure within identity does not permit. The individual, in losing himself, in expending herself in such a 'society of souls', knows or feels that he or she is gaining a 'boost of existence', that of participating in a community that turns his or her loss into a net profit.

(Maffesoli, 1996b: 58)

INTRODUCTION

In this final stage of the night, I develop what has been suggested throughout 'The night out' so far in respect of crowds, music, dancing and the clearly important roles of the ecstatic and oceanic experiences in clubbing. Fusing these different facets of clubbing, I sketch out a conception of the relationships between notions of power and resistance that problematises and gives texture to certain existing conceptions in which resistance is clearly directed outwardly and around apparently straight-forward conceptions of authority and subordination. The alternate conception that I introduce in this last section of 'The night out' is one that I am calling 'playful vitality'.

This is a conception of the vitality that can be experienced through play, and I develop it in more detail through three main sections. First, after noting a general lack of intellectual engagement with the world of play, I suggest clubbing is a form of play through which a sensation of 'flow' – a matching of challenges with skills and techniques – may be experienced. Clubbing as a form of play can, I propose, be inwardly rewarding in a consequently often subtle form. Second, I develop this notion of clubbing as playful in critiquing and extending some existing notions of resistance and domination that are currently *en vogue*. Specifically, I suggest that certain form or conceptions of 'power' can also be understood as 'vitality'. That is, vitality can be immanent and inwardly oriented, contrasted with and against other perhaps less pleasurable aspects of an individual's life. Third, I relate playful vitality more explicitly to clubbing through highlighting this autotelic quality within the practices and spacings of clubbing, but also simultaneously stressing its vital group context.

I conclude by problematising what seem to be sometimes over-romanticised and idealistic visions of the supposed annihilation of social differences upon the dance floor. These visions, often invoking notions such as 'the gathering of the tribes', would appear to be founded more within what may be termed 'sensibilities' than 'realities'. However, this recognition of the imaginative quality of certain facets of clubbing does not deny the crucial value to many clubbers' experiences of these emotional–imaginative understandings.

PLAY AND FLOW

I do not think it sensible to ignore, as most rationalists have done, ecstatic experiences and the emotions or ideas to which they give rise. To ignore or to deny the importance of ecstatic experiences is to leave to the irrational the interpretation of what many people believe to be of supreme value.

(Laski, 1961: 373)

– Just what is it that you want to do?
– We want to be free...we want to be free to...to do
what we want to do!
– And we want to get loaded, and we want to have a good time.
– And that's what we're going to do...we're going to have a
good time, we're going to have a party!

(Primal Scream, 1991, 'Loaded')

BEN: What do you think about at the club?

LUKE: Everything and nothing...Sometimes my head really seems to be empty, but I can have many thoughts as well. If I think, it's a bit in the background and my thoughts are a bit like when you're tripping [using LSD] though no drugs are involved. Many little thoughts, just popping up and dissolving. Sometimes I think of the things I have or had to do and my problems...but not in a nasty way, it's just like I organise things in my head, maybe the sort of dreams you have when half-asleep. Sometimes, I think about the people around me. That may be because I like someone or because I don't like someone. If someone pays me attention for some reason, seems to be threatening, talks too much in an offensive way or whatever, that can annoy me. Or

135

I think about the girls around me. How pretty they are, or how to get close to them. Or I think about myself, about how good I feel. And, sometimes I think about how I look. It's not something I think about the whole night, but I am aware of how I dance, and that I might be 'cool' or 'good looking' [uh-huh]. Don't take that all too literally though! In brief, it's a bit hard to tell you what exactly I think about. It depends on my mood and what's happening around me. Surprisingly, drugs don't really seem to influence what I think about. More indirectly, though, E normally makes me feel good, so I have good thoughts...ehmmm...fairly...E makes my head rather empty. Maybe I dream a bit more and have more freaky thoughts.

The subtle yet important experience of dancing for Luke is evident in this extract. Luke's head is both 'empty' – for that brief moment he does not have to or need to think – yet his perceptions of himself, of those around him, about the experience of which he is a part and of his life beyond it clearly dominate his thoughts – 'sometimes I think of the things I have or had to do and my problems'. For Luke the dance floor and the experience of dancing is both a space of dreamy non-thinking and a space of self and situational monitoring of an intensive quality – a space of play.

Play

It is perhaps obvious to clubbers that clubbing is a form of play, yet at the same time more than a meaningless way of passing time. However, in academic terms play and playing have often been neglected as strands of social life. Play is seen 'not to fit in anywhere in particular' (Turner, 1983: 233). Play is seen as a 'rotten category...tainted by inconsequentiality' (Schechner, 1993: 27; cited in Thrift, 1997). Yet play matters, for it is central to our everyday lives. To play is to spend time, exert energy and employ techniques on an activity that can be self-contained (or autotelic), cathartic and, at times, ego-expressive (Giddens, 1964). Play is a voluntary activity that is positioned in some way as opposed to 'work', both in terms of its location and duration – play contains its own course and meaning (Huizinga, 1969) – but also in terms of its content and practices. Play is different through its definition as 'play', yet like non-play, play also has rules and limits (Csikszentmihalyi, 1975b). Playing can be a source of identity and even of a form of power, but play does not necessarily come easily. In a similar way to that in which work can be playful, play often needs to be worked at. Thus, the same activity can be play to one person while another experiences it as work – various sports for example. Csikszentmihalyi (1975a) even suggests that for some surgeons the practices of surgery itself can be playful (this is explained further on p. 139).

Approaches to understanding or constructing a role for play have variously positioned it as a long-range survival technique in that it prepares young people for adult tasks, as an outlet for unexpressed needs and as compensation for routine behaviour (Young, 1997). These are explanations that have broadly concentrated upon adaptation to a changing environment and associated changes in responsibilities and expected behaviour (Csikszentmihalyi, 1975c). Furthermore, play has been proposed as being the result of a physiological need for optimal arousal. That is, play is valuable in that it provides stimulation in parts of the brain that are not utilised in other human activities such as work or sleep (Csikszentmihalyi, 1975c). Play has been posited as being about stepping out of ordinary or everyday life (Huizinga, 1969; Young, 1997), while for others 'man [sic] only plays when in the full meaning of the word he is a man, and he is only completely a man when he plays' (Schiller, 1967; cited in Young, 1997: 75). While the slightly obtuse language of Schiller's point can be more easily understood when it is noted that the original text was first published in 1795, the point he makes is as relevant today. One may often feel more human when one plays. That is, to feel you are 'working' or have to 'work' at a certain time is not to be free, not to be – in a sense which is difficult to capture – a real 'human'.

Thrift (1997) and Bauman (1990; 1993) place less significance in play than do Csikszentmihalyi (1975c) or Huizinga (1969). Zygmunt Bauman, for example, understands play as 'gratuitous...serving no sensible purpose' and which, when called upon to reveal its function, displays its 'utter and irremediable *redundancy*' (Bauman, 1993: 170; emphasis in original). I disagree, instead understanding play as potentially refreshing and revitalising, and as performing an important role in (to give but two examples) social and sexual interaction, neither of which I would suggest were 'redundant' aspects of social life. Neither is play 'free', as Bauman (1993) asserts. Play demands of the individual 'player' at least as complex a set of techniques and competencies as the worlds of work. The hierarchies of power are often less clear in play – the 'rules of the game' more open and up-for-grabs – and those playing together are more likely to include some who have not before played in that social context. When play is explicitly a group activity, as in team games and crowd-based events and situations, the 'rules' of the game become even more prominent and the associated competencies much more valued as the additional notions of acceptance, belonging and identification explicitly impact upon the situation. Like 'work', play involves knowledges of 'ways of doing', conventions, customs and competency at their timely implementation. Play thus involves not an absence of orderings, but a set of *alternate* orderings (Hetherington, 1997).

Bauman further suggests that it is the sense of being 'free and gratuitous...that sets play apart from a "normal", "ordinary", "proper", "real" life', and thus play is 'not for real' (1993: 170). Yet I would suggest that play is merely another facet of an everyday or 'normal' life, that play exists not outside, but *as a constitutive part of* our everyday lives. Certainly, play differs qualitatively from other facets of everyday life such as 'work', but play remains a significant facet of that everyday life and not something optional and outside experience.

Following this, in suggesting that play 'does not add up' – that, in other words, play has no direction or continuity outside of the moment of playing – Nigel Thrift (1997) perhaps overlooks the critical roles afforded by many individuals to play, playful times and spaces, and play-based activities and experiences. Play is an intrinsic part of our everyday lives, not only affecting but, in many cases, also partially constituting central aspects of our identities and identifications – aspects that we carry through our non-playing times and spaces. Time may take on a different quality and feeling during the practices of play – 'time flies when you're having fun' – and the spaces and contexts of play might be perceived as different or special, and endowed with a powerful significance or even totemic quality. Yet this sense of being different has meaning only inasmuch as play is a part of, and exists only in contra-distinction to, the realm of non-play. In any case, play does (to use Thrift's term) 'add up' in the important sense that individuals can get 'better' at playing.

Turning to the dancing that is so central to clubbing, it is clear that clubbers feel 'better' or more competent at dancing after some time practising, and that as they feel 'better' at dancing so their enjoyment of that activity increases. This is a point made forcefully by both Seb and Sun in the discussion of dancing (pp. 88–92, Section Two of 'The night out'). With practice, play requires less 'work'. As sensations of competency at playing increase, a sense of 'flow' becomes possible.

Play as flow

Do activities for which extrinsic rewards are minimal provide a set of intrinsic rewards of their own; if so, what are these intrinsic rewards?

(Csikszentmihalyi, 1975c: 179)

SEB'S JOURNAL ENTRY: If ever there was a call-to-party this was it. For once, it was our turn to look around in bewilderment. Everybody could see what was taking place – we were all part of it, yet we all gasped in wonder. It was truly astonishing to witness. (Even I was tempted to ask 'What have we started?'…but I stopped doing that ages ago!) To 'go with the flow' is the accepted response to this captivating, hedonistic magic that *is* the Party Posse and which follows us wherever we go. That's all there is to do. Tonight the wave which swept us along was as sweet as honey: a radiant mixture of all the hopes, vitality, passion and enthusiasm shared by an extraordinary bunch of like-minded party people. The energy generated at that one instant as I hugged Noel and grinned at all those others glowing around me was fucking electric. We were charged for a party and as it turned out we proved ourselves very worthy![57]

In this wonderfully evocative, almost poetic, description of the experience of the dance floor during the climax of the evening, there are captured a number of important points about play and about the sense of flow that can accompany it. First, there is a palpable sense of disbelief amongst Seb and his friends about what they were actually part of and were witnessing, as, 'for once', it was their turn to be astonished. Second, what they were witnessing had 'called them to party' – the experience of the euphoric and ecstatic crowd had forged within them an even deeper sense of ecstasy and euphoria – it was a self-magnifying process. Third, Seb himself describes the experience as one of 'going with the flow'; thus my employment of Csikszentmihalyi's (1975c) term 'flow' does have some foundation within the experiences of clubbing themselves, although perhaps unsurprisingly Csikszentmihalyi does not mention clubbing. Fourth, Seb's use of the metaphor of the 'wave' that 'swept them along' is reminiscent of the clubbers' descriptions of experiencing the oceanic that I presented in the previous section of 'The night out' – 'a radiant mixture of all the hopes, vitality, passion and enthusiasm shared by an extraordinary bunch of like-minded party people'. Finally, throughout this extract, but also the journals of Seb's experiences as a whole, there is an emphasis on the 'we', on the group and its strong identification and tightly-bound constitution. Seb rarely talks in terms of his own personal experiences, instead repeatedly referring to 'the Party Posse', 'we' and 'our'. Seb is recounting a shared experience, as well as one that is obviously extremely rewarding on an individual level.

In his elaboration of a 'new theoretical emphasis for the understanding of human motivation' Mihalyi Csikszentmihalyi (1975a) sets out to problematise the conventional distinctions between work and play. His main theoretical tenet is based upon the notion that a 'sense of flow' can be experienced from the meeting of challenges with learnt or acquired skills and competencies. Any activity in which this sense of flow is experienced (work-based, play-based or a blend of the two) can be understood as a pleasurable activity (Sadgrove, 1997), and the quality of everyday life depends partly upon the balance between the challenges experienced and skills required in certain situations (Moneta and Csikszentmihalyi, 1996). These flow experiences, argues Csikszentmihalyi, indicate that (at least in his terms) it is less the label or category of 'play' itself that determines the pleasure inherent within or potentially gained from an activity, and more the timely and embodied practice of skills in the negotiation of the various challenges that a given activity – work and/or play – offers to an individual (Csikszentmihalyi and

Rathunde, 1993). This point may be clearly illustrated through the example of dancing. Whether one dances for *pleasure* – for example, as a clubber, in carnival, or as part of a line dancing, contra-dance or ballroom dancing group (see Levine, 1987) – or as a form of *work* – for example, as a ballerina, as a 'dancing' sex worker (see Law, 1997) or stripper, or as a cheerleader at a sports event – the demands for competency and the pleasure experienced through meeting the challenges posed by the social situation through that competency override the polarised notion of dancing simply as being either 'fun' or as 'work'. It follows that whether dancing is enjoyable or not depends not only upon the fact that it can be a leisure pursuit and an occupation:

> Apparently, something besides the activity itself must be analysed to decide whether it is intrinsically rewarding; specifically, one must consider the structure of external rewards that ties the activity to other social institutions.
>
> (Csikszentmihalyi, 1975c: 180)

Like Csikszentmihalyi (1975a), I see play as significant in understanding the social lives of individuals and groups both for its *intrinsic* value, that is, in terms of the person's subjective experience of play and playing, but also for its *extrinsic-value* – its relationships with the ways in which individuals and groups function and live their lives outside of the specific contexts of that play. Within any group situation involving practices of play, such as clubbing, there will be variation in the extent that play is significant intrinsically and extrinsically, and this variation will exist between individuals and also within the experiences of a single person throughout the timings and across the spacings of that situation.

Whether and to what extent clubbers experience dancing as an intrinsically and/or extrinsically rewarding experience might be understood in terms of whether they experience the clubbing experience as what Csikszentmihalyi refers to as a 'flow experience' or as an 'experience of boredom and anxiety'. The former is used to refer to an experience in which skills and competencies are successfully employed in overcoming or meeting understood challenges, and the latter is used to refer to an experience where there are either *too many* or *too few* opportunities for action – the individual is either over-skilled or under-skilled to cope.

Through this 'flow lens', clubbing, and the techniques and spacings of dancing in particular, can be understood as foregrounding the matching of personal skills, acquired through mimicry, observation and practice, with a broad and constantly changing range of physical or symbolic opportunities for action, where these opportunities represent 'meaningful challenges to the individual' (Csikszentmihalyi, 1975b: 102). Thus, the oceanic and ecstatic experiences, which I discussed in the previous stage of 'The night out', and of which Seb's journal entry (pp. 138–9) is powerfully evocative, might be understood as moments of *extreme* flow for the clubbers. These transitory sensations of extreme flow can be experienced through the meeting of additional emotional and identificatory challenges – challenges which I discussed in my unpacking of the emotional spacings of dancing – in addition to those of dancing techniques, or crowd interaction, and perhaps of sexual interaction and the consuming of drugs. During these extreme flow experiences, of which the oceanic and ecstatic in clubbing are just two examples, those 'going with the flow' might:

> [C]oncentrate their attention on a limited stimulus field, forget personal problems, lose their sense of time and of themselves, feel competent and in control, and have a sense of harmony and union with their surroundings...some people emphasize the movement of their bodies; others try to maximize emotional communication; still others respond to the social dimensions of the activity...
>
> (Csikszentmihalyi, 1975c: 182–3)

SEB'S JOURNAL ENTRY: With The Main Room packed, Jon of the Pleased Wimmin [a DJ] loaded an anthem that had 800 sets of hands, 3 brussel sprouts and 1 decapitated head in the air (bearing in mind I was tripping, don't ask too many questions). From our vantage point just in front of the DJ box, myself and those around me could envision the emotion flowing across the floor and through the already entranced crowd of party people. This sight only served to heighten my own rush as I became overwhelmed in ecstasy...delirious and delighted because of it.

As the night wore on, I became oblivious of time, only conscious that I was having a great one.

And everybody else...well, their collective state was pretty much *'fucked'!* – but on everybody's face was a story of an exceptional party night.

Seb evokes the self-reinforcing notion of the flow experience that can be experienced through dancing. The sight of the 'entranced crowd' leads Seb to become 'overwhelmed in ecstasy'. In this case, Seb is overwhelmed both through the combined effects of ecstasy (MDMA) and the physical sensation – *exstasis* – produced through the crowd, the music and the movement. Seb's careful noting of the temporary disappearance of time – once again, a concentration on the 'here and now' – and his referring to the dancing crowd as one ('everybody else'...'their collective state'...) are characteristic of both the experience of flow and also of the ecstatic experience. This is a social situation seemingly beyond

141

regular times and spaces. This sensation of 'being controlled' by the situation, of being carried along with it, is redolent of Valerie's description of her experiences of the dance floor which I mentioned during my discussion of dancing – 'I was FORCED to do it, what I was doing I was being forced to do by something'. This particular experience can now be understood as both a form of oceanic experience, but also as being constituted through sensations of flow. Furthermore, dancing can in this way be understood as a playful activity through which flow, at times of overwhelming proportions, may be experienced. Dancing is a playful activity through which the relationships between self and those that constitute the surrounding dancing crowd are, at the least, unsettled and in certain cases temporarily re-cast completely.

Maria Pini refers to the work of Haraway (1991) and Braidotti (1994) in her elaboration of how, in 'rave', the physical body of the woman comes to include technology 'in the forms of music, lighting and drugs' and is perhaps better understood 'in terms of a mind/body/technology assemblage' (Pini, 1997a: 124). Thus, arguably, Valerie could be understood in this way as becoming a 'Cyborg' (Haraway, 1991), with her body being taken over or 'controlled' by technology (Pini, 1997a). Whilst fascinating, these points seem to spin more off the (feminist) agenda of the theorists concerned than the experiences of the raving women being discussed. The male clubbers whom I interviewed evoked their experiences in a similar fashion – take Seb for example – and in any case I would place more emphasis upon the flux between *self and the clubbing crowd* than upon any flux between human and human technology assemblage (I further discuss these points on p. 151).

That said, it should be added that the flow experience works both ways. The experiencing of flow through play is by no means automatic, and clubbing can also be an experience laden with anxiety. The demands of the social situation can become too much; for example, in instances of clubbers new to a certain club, scene, or even to clubbing itself, or perhaps at certain moments of the night. How many clubbers, however 'cool', experience 'flow' in the queue? This anxiety arises as a result of the general management of what a clubber might describe as a 'cool identity'; that is, the presentation of self in certain stylish ways so as to become accepted, and ideally accepted as 'cool', through the manipulation of body techniques, dress and physical appearance. Anxiety can thus arise because of a sense of complete *non-flow*, the sense that one does not belong, or does not wish to belong, or perhaps does not know how to begin to establish belongings even if one so desired.

BEN: Do you do a lot of thinking when you're out?
MARIA: It depends on the evening I think, ummm…there are occasions where, errr, you're very relaxed and you don't really think that you're too heavily

enjoying yourself, um, if that makes sense, erm, not really!

BEN: So you're quite relaxed, or...?

MARIA: Well, yeah, I mean if I think I'm comfortable in a situation...to put it into context erm...I recently, I'm seeing one of these people that was organising Saturday – I only met him three months ago. Went down to Bristol a while back to see him and he wanted to go out to a club night at Freeform, it's a...it was a non-Freeform night basically, taken over by sort of free party people, and it was the weekend before Tribal Gathering and it was actually a very bad turnout and I only knew him, and I've met his flat-mate very briefly and I was in a sort of...I dunno...[yeah?]...not having a clue how to get home if I didn't enjoy myself, with two people...well...umm...he went off on a mad one, just running around for like an hour – I didn't see him for ages. And it probably took me an hour and a half to adjust to my new surroundings, so in those sort of situations I would say yeah, I definitely...I do think a lot more, in other situations if I'm in a club, if I know a few people down there then it's...y'know.

In this short dialogue Maria indicates how clubbing may lead to an anxious experience rather than one of flow. The experience that Maria narrates is certainly unlike the oceanic and ecstatic experiences which I outlined in the preceding section and to which Seb and Valerie allude (pp. 138 and 133 respectively). Moreover, the very fact that Maria mentioned this night as memorable despite the obvious lack of fun that she experienced suggests the extent of her discomfort – it certainly left its mark upon her.

Clubbing, then, can provide opportunities for elements of flow to be experienced. That is, clubbing can have intrinsic value and can allow a matching of challenges with skills and techniques. However, clubbing can provide instances in which flow is inhibited. In particular, a dancer may be interrupted in their dancing practices, they may experience a 'break in the flow' and become self-conscious. A dancer may even get ambiguous or unpleasant feedback from those within the crowd. Thus the 'edges' of the dance floor can be regionalisations of increased self-consciousness because of the presence of non-dancing clubbers who appear to be watching those dancing. Yet, for similar reasons, the edges of dance floors can also provide opportunities for overtly staged displays of skill, dancing styles and self-control – a static audience is also a captive audience (Csikszentmihalyi, 1975b).

The demonstration of flow whilst dancing – of apparently coping successfully with the demands of expressing bodily understandings of music and the social situation – is thus an important element in establishing oneself as competent, as 'cool', and as clearly belonging. Dancing 'well' is not only important for the intrinsic rewards it provides for the clubber, but is also important as it forms one of the major character attributes (Goffman, 1968) through which clubbers are understood socially by others within the clubbing crowd.

As with in any other form of social situation, the challenges of playing are experienced in practical terms in the guise of the rules, customs, standards and traditions through which that play is constituted. Yet, while play may well be 'rule-bound' (Thrift, 1997: 146), these rules – the customs, 'ways of doing', practices and techniques constituting the sociality of clubbing – overlap into, and themselves partially constitute, other aspects of the clubbers' everyday lives. Play can itself be a source of identity and of identifications, and can instil notions of community and belonging even outside the time–space of that playing. The 'rules' of play are important – they do partially structure play – but they also go *beyond* play. Notions of competency and identification extend well beyond the moments of flow themselves and into the individual's sense of self identity.

Furthermore, while play, in this instance in the form of dancing, may serve 'no useful purpose' (Thrift, 1997: 145–6) in being beyond rational notions of 'functionality' on the one hand, on the other hand play may be an important source of strength or vitality for the individual dancers, both in terms of their individual identities (how they see themselves) and their identifications (with whom and how they identify – their belongings). This vitality can be drawn by the clubbers through the experiencing of flow. In extreme cases this flow is experienced in the form of moments of oceanic or ecstatic euphoria. The paradoxical in-betweeness of these moments – fleeting sensations of both losing control yet also finding control – can result in sensations of *exstasis*, of the dancer losing a sense of him or herself as a separate entity, of becoming part of or identifying strongly with something outside and beyond, yet also including, themselves.

In the remainder of this final section of 'The night out' I want to build upon these notions of play and flow through making two further points. First, I want to re-visit notions of resistance in sketching out the conception of resistance as 'vitality'. Second, I want to blend aspects of play and flow that I have just outlined with these notions of vitality and, both through the words of the clubbers themselves and through critiquing existing accounts of this form of 'other-world', sketch out the notion of 'playful vitality' more fully.

RESISTANCE AND VITALITY

Club-culture [sic] remains unashamedly escapist. The hedonist motto, 'Don't Worry, Be Happy' is undoubtedly very popular. Ignorance is Ecstasy. This indifference to the world of politics can lead to a subjectivity where criticism plays no role.

(Ali, 1998: 15)

What if, instead, we were to consider 'resistance' in other less fundamental ways?

(Thrift, 1997: 124)

[P]eople are positioned differently in unequal and multiple power relationships...more and less powerful people are active in the constitution of
unfolding relationships of authority, meaning and identity...these activities are
contingent, ambiguous and awkwardly situated, but...resistance seeks to
occupy, deploy and create alternative spatialities from those defined through
oppression and exploitation. From this perspective, assumptions about the
domination/resistance couplet become questionable.

(Pile, 1997: 2–3)

There has been an increasing impetus across the social sciences over the last thirty
years to illuminate the words and worlds of the so-called 'powerless and dominated' in societies (Hetherington, 1997). In these studies, the relationships
between the so-called 'powerful' and the 'subordinated' have often been
constructed in terms of the worlds of work and/or as relationships in which the
weak are exploited by the powerful and thus exercise tactics of resistance in their
relationships with those in power (Scott, 1985; 1990). Others have broadened
this focus upon resistant practices to include the ways in which, for example,
clothing, food and writing can be subject to tactical subversions (de Certeau,
1980; 1984), and have questioned the extent to which (to give another oft-cited
example) the everyday practices of walking in the city coincide with the intentions and orderings of the city planners (de Certeau, 1993). Still others have
examined the inversion of the everyday that can occur during carnivals and fêtes –
Scott calls these practices 'rituals of reversal, satire, parody, and a general suspension of social constraints' (Scott, 1990: 173).

In discussions of 'youth' or young people as well, resistance has often been
variously cast in terms of a defiant, stylistic insubordination (Chambers, 1986;
Hall, 1984), as an attitude or ritual of resistance to those in power or in authority
(Hall and Jefferson, 1993; McKay, 1996) and as forming a process of 'bricolage'
and symbolic transformation (Hebdige, 1979; 1983; 1988). Throughout, there
has been a powerfully resonant notion of a power, a hegemony or an authority
against or in defiance of which these young people have been acting.

More recently, in certain feminist and cultural geography perspectives on relations between spaces, the nature of power, and identities, issues of domination
and resistance have started to become more textured and problematising in their
approach (Hetherington, 1997; Keith and Pile, 1993)[58]. I want to build upon
some of these recent approaches, and specifically upon the suggestion that it is no
longer enough 'to begin stories of resistance with stories of so-called power',
with the process of thinking about and attempting to broaden understandings of

'geographies of resistance involving breaking assumptions as to what constitutes resistance' (Pile, 1997: 3).

In constructing clubbing as simultaneously a form of play and as constituting a form of vitality, I want to highlight a differing inflection of 'resistance' and of 'power'. I am interested in resistance less in terms of being concerned with effecting changes in a macro or overtly stated context (or scale), but rather as constituting a situation in which an 'alternative conception of the self' may be fostered (Pini, 1997a: 118) – an alternative conception which may provide a sense and a source of vitality, or personal worth. This situation is produced through a temporary replacement of the 'normal' and normative codes and customs of interaction and communality with a set of alternate orderings. Before I turn to this in detail and attempt to contextualise my argument through the stories of the clubbers, I first want to draw out a number of key pointers from some existing discussions of the relationships between power and resistance.

Power and resistance

In one of the most influential discussions of the domination–resistance relationship in recent years, Michel de Certeau (1984) uses what is by now a set of well-rehearsed arguments in noting how the practices of resistance form a popular culture that, through continual evolution and mutation, resists 'assimilation' by the powerful and the elite. This popular resistance to situations of domination and the imposition of power is neatly illustrated by de Certeau through his use of the notion of *la perruque* (or 'the wig'). He outlines, for example, how in some cases a worker's own 'work' – perhaps talking to friends via the Internet or reading, whilst ostensibly working – may be disguised as productive work to his or her employer:

> In the very place where the machine he [sic] must serve reigns supreme, he cunningly takes pleasure in finding a way to create gratuitous products whose sole purpose is to signify his own capabilities through his work and to confirm his solidarity with other workers or his family through spending his time in this way.
>
> (de Certeau, 1984: 25–6)

James Scott (1985; 1990) develops a perspective on resistance and domination which is strikingly similar to that of de Certeau, yet seems to offer a more contextual and in some ways more approachable conception than de Certeau's polarised 'strategies' and 'tactics'[59]. Although largely basing his arguments on anthropological work carried out in a small Malay village, Scott nevertheless offers a cross-cultural perspective in proposing that, other things being equal, structures of domination can be demonstrated to elicit patterns of resistance which are comparable in form. That said, Scott is quick to recognise how the 'ultimate value of the broad patterns...could be established only by embedding them firmly in settings that are historically grounded and culturally specific' (ibid.: xi).

Like de Certeau, Scott suggests that *open* resistance to the practices of domination is rare – a point he makes in relation to the slaves and serfs in the villages that he studied, yet that could be equally applied to, say, workers in the City of London, bar staff in a nightclub or someone using a 'second-hand' ticket in a pay-and-display car park. Every subordinate group fashions a 'hidden transcript' out of its ordeal – an embodied and spatialised critique of power that is almost always spoken behind the back of the powerful (Scott, 1990). Thus, much more likely than open resistance is an opposition to domination or the powerful that is spatialised, occurring within specific places, and clandestine[60]. The 'hidden transcript' comprises the 'off-stage' or backstage responses that are carefully choreographed in spatio-temporal terms (Scott, 1990: 109).

Both the accounts of de Certeau and of Scott, which I have only briefly mentioned, privilege the realms of overtly politicised resistance. Even where this resistance is clandestine in nature it is 'aimed' at simple notions of 'power', albeit sometimes unintentionally. Both sets of accounts are concerned with apparently dualistic constructions of 'power' versus 'resistance', in which the less powerful resist or subvert the apparent control of the more powerful through their tactical responses to the strategic control of the powerful, whether this is through re-inscribing the names of streets with different and unintended meanings (de Certeau, 1984) or through the singing by slaves of thinly veiled (but still 'hidden') attacks in the form of hymns on slave owners (Scott, 1990). I want to look at an alternative facet of this power–resistance relationship.

Play as vitality

Understood in one way, play can be about using the body to 'conjure up "virtual", "as-if" worlds by configuring alternative ways of being through play, ways of being which can become claims to "something more"' (Thrift, 1997: 147). Play, whether involving nine holes of golf, an hour in the kitchen with a cookbook or a night spent on the dance floor with friends, is about the temporary inhabitation, and thus constitution, of alternate worlds and their orderings (Hetherington, 1997) in which rules that are different from those dominant within 'normal' life usually apply and through which, in some way, a sense of vitality, of personal worth, of energy and of reward is experienced (Csikszentmihalyi, 1975c).

If this engagement with apparently 'different worlds' or imaginary realms that can characterise play is about the eluding of certain forms of macro-power and structurings in the social lives of actors – such as in work, the demands of sociability, the legal system, the class system, the categorisation of the self – as Thrift (1997) contends, then it is not *only* about this. Play is also about an *engagement with* and an *expression of* a different facet of power altogether. This power comes not from above – it is not ascribed – but from within – it is achieved. Rather than being a mode of power that is *evaded* through play, it is instead a form of micro-power or 'vitality' that can be *inhabited* through play.

JOHN: For me it's not like escaping from a particular place, like, geographi-
cally…it's about escaping full-stop. Just escaping from everything I think.
Escaping from work, escaping from normal life, escaping from everything,
everything full-stop. Ummm…it's almost as if…I don't know, you step into a
totally different life – nothing at all…nothing else matters [no] and once
you're at the club, or on your way to the club or whatever, or on a night out,
whenever it starts – whether it's six o'clock when you leave work or eleven
o'clock when you walk into the club [umm], that whole night, and the next
morning when you're sitting around indoors chilling out and smoking,
is…just going to be my time, our time [yeah, yeah]. No one else's wants or
needs or anything comes into it, I mean the other people I'm with – yeah
[ummm], I mean them having a good time or…I might give up a little time
to make sure of them, but I'm just escaping from everything else.

BEN: How do you feel when you're actually there then, at the club?

JOHN: Cut off a bit, safe…in some respects I don't even think about it, I don't
worry, that's pretty much it, it's not so much that I don't worry about this or
I don't worry about that, or I put this to the back of my mind, I don't need to
put anything to the back of my mind really. You don't think, you just do
[yeah]. And that is the best feeling about it – you just don't care about
anything – you just enjoy yourself!

The subtle nature of the vitality that John experiences through and while clubbing is tangible in this delicately descriptive extract. John repeatedly uses the
term 'escape' – he's escaping not from a place so much as from 'work', from
'normal life', from 'everything full-stop'. The clubbing experience is clearly
different from other aspects of John's everyday life. Furthermore, whilst clubbing, nothing beyond or outside it matters – 'this is just going to be my time, our
time', he says – a time quite unlike that of work and of life beyond the here and
now. The obligations of thinking about his actions and his worries are also
temporarily suspended while clubbing – 'you don't think, you just do' – and there
is a clear sense in what John says of the experience being rewarding for reasons
deeper than simply being just 'fun'. This is certainly no crude hedonism that John
is evoking.

Following John's story, it is possible to see how we might begin to understand the practices of sociality that constitute play more as implicit notions of autotelic *strength* or *vitality* than of power or resistance in the face of any explicit, dominating authority. In this sense play can be understood not so much as a reaction *to* 'all the dangers and dysfunctions of the moment' (Maffesoli, 1996b: 48), but more as *in spite of* them:

> It is certain that unemployment, violence, economic constraints, the threats of moralism, and other forms of alienation are felt as so many impositions that bridle, alter, or hamper a flourishing social and individual life. But all that does not prevent one from being engaged in experiencing, at best, what one can enjoy. One could even say that, faced with these constraints, there is a frenzy to enjoy, a *carpe diem*[61], a surfeit of social energy that no longer is carried over into the future, but invests itself in the present.
>
> (Maffesoli, 1996b: 48)

Through the matching and superseding of often self-posed challenges with acquired skills and competencies individuals can experience sensations of control over their body, can re-invigorate their sense of self-worth, eliminate alienation and experience an exhilarating feeling of vitality (Csikszentmihalyi, 1975c). Michel Maffesoli calls this form of micro-power or vitality *puissance*[62], suggesting that it is a form of 'underground sociality' that can be expressed and experienced by individuals through 'trivial vectors' such as places of everyday chit-chat and conviviality (1995: 3). For Maffesoli, *puissance* is the 'will to live' and it 'nourishes the social body' (ibid.: 31). Practised through 'abstention, silence and ruse, puissance is the opposite of politico-economic power' (ibid.: 3). It is derived from the 'proxemic reality of community' (ibid.: 31) rather than from any sense of counter-authority. Unfortunately, while this distinction between 'tactile' and 'optical' understandings of others is useful, Maffesoli typically fails to develop his notion of *puissance* in a more textured and useful fashion, instead merely suggesting little more than just that 'this energy is difficult to explain' (ibid.: 32).

Building upon the sentiment underlying Maffesoli's conception, and through re-positioning play as a form of activity through which flow may be experienced and a sense of vitality (re-)forged, I propose that apparently autotelic and playful practices such as dancing may be at the very centre of many individuals' attempts to make themselves significant both to themselves and also within group social interactions. However, before I move on to discuss this notion of playful vitality in detail, I first want to further refine exactly how clubbing might be understood as potentially providing vitality to clubbers.

Now *everything* is 'resistance'?

I am aware that I am not alone in these explorations into the nature of resistance. In the wake of the strategies and tactics of de Certeau (1984), in particular, has come a tidal wave of discussions and debates about resistance as being present in the minutiae of individuals' everyday lives. Conscious of the rather narrow conflation between dualistic notions of resistant practice and powerful, subordinate relationships premised either within the worlds of working or overly structuralist notions of 'subculture', and based upon clear constructions of 'authority' or 'surveillance', many have recently attempted to broaden their focus. This flood of reconfigurations has reached such an extent that Pile is driven to wonder: if resistance 'can be found in the tiniest act – a single look, a scratch on a desk – then how is it to be recognised as a distinctive practice?' (1997: 15); and Morris mockingly suggests that one can discover 'washing your car on a Sunday is a revolutionary event' (1988: 214), while for Cresswell it seems that 'almost any activity from eating to walking to writing books and making films can...be construed as resistance' (1996: 22). In further qualifying my arguments about clubbing and playful vitality, and in order to avoid accusations of 'romanticization' and 'misguided optimism' (Cresswell, 1996: 22) in discussing certain practices of play as forms of resistance, I want to set out in some detail how and why, in the practices and contexts of certain playful activities, there exists the realisable (and realised) potential for the experiencing of personal and social vitality.

First, I am arguing quite simply that play can be experienced as a form of resistance and an oppositional practice in the guise of what I have described as 'playful vitality'. By this I mean a reconfigured notion of resistance that falls somewhere between the purposeful and outwardly orientated, macro-politicised sociality of Scott and de Certeau, the immanent and, at times somewhat mystical, sociality of Maffesoli – who never pins down exactly who it is that is supposed to be experiencing strong feelings of *puissance*, and how and where this is directed – and the explicitly technical sociality inherent within Csikszentmihalyi's notion of 'flow'. Playful vitality is purposeful and directed, yet inwardly orientated. Playful vitality privileges sensations of flow, of the here and now and of the tactility of the crowd.

Second, and related to this, I am suggesting that de Certeau's *la perruque* might be a notion that we can expand and apply not just to situations of *individualised* tactical responses in both institutionalised (work or production) and consuming contexts or settings (walking, reading, cooking), but also to *crowd-based* social situations, many of which are based upon the practices of playing and having fun as a group or crowd, such as carnival, and involve notions of 'power' which are not directly related to conventional constructions of 'authority'. As Steve Pile proposes:

> [R]esistant political subjectivities are constituted through positions taken up not only in relation to authority – which may well leave people in awkward, ambivalent, down-right contradictory and dangerous places – but also through experiences which are not so quickly labelled 'power', such as desire and anger, capacity and ability, happiness and fear, dreaming and forgetting.
>
> (Pile, 1997: 3)

Thus, and thirdly, I am proposing that through play individuals may offset or counterbalance other less playful and less flow-based, and thus less rewarding, aspects of *their own lives* in both working and playing contexts. That is, rather than 'resisting' notions of power which are acting upon them, individuals may actually be 'resisting' other facets of *their own* identities, certain of their other identifications and even some of their other emotional experiences. Importantly, it appears that this form of 'self-resistance' might be easiest to experience in group situations.

Fourth, I want to move beyond simplistic conceptions of alternate orderings that involve and attempt to reify certain social encounters as in some way (even) temporarily obscuring, incorporating or annihilating differences in terms of the identities of those taking part in that encounter. In the final part of this section I argue that conceptions such as Turner's (1969; 1982) *communitas* idealise the inclusivity of 'differences' that occurs during moments of group effervescence. While not denying the genuine nature of the *sentiments* of inclusivity that can occur in clubbing, I propose that the *actual* differences of those within clubbing crowds should be carefully scrutinised. I now turn to look at the relationships between clubbing and the notion of playful vitality.

CLUBBING AS PLAYFUL VITALITY

> Music creates order out of chaos; for rhythm imposes unanimity upon the divergent; melody imposes continuity upon the disjointed, and harmony imposes compatibility upon the incongruous.
>
> (Menuhin, 1972: 9)

> We are together,
> We are unified and of one accord.
> Because together we got power,
> Apart – we got pow-wow.
> Unified, we are together.
>
> (Primal Scream, 1991, 'Come Together')

It is for the sake of this blessed moment, when no one is greater or better than another, that people become a crowd.

(Canetti, 1973: 19)

Moving back towards the clubbing experience in contextualising this concept of playful vitality, I make three related points. First, I highlight the autotelic nature of clubbing in noting its lack of explicit outwardly oriented politics. Second, through drawing out the unburdening of individuality that can be experienced during the clubbing experience, I note how playful vitality is very much collectively experienced as well as individually valued. Third (and in a little more detail), in attempting to defuse, problematise and qualify the romanticism and idealism that often exists around discussions of communality and difference in respect of clubbing, I suggest that sentiments of openness and inclusivity are often misconstrued as suggesting a hugely diverse collection of identities on the dance floor when this may well not be the case. However, this is not to deny the validity and authenticity to the clubbers of these experiences of a 'better' or 'ideal' world in which relief from the mundaneness and tensions of other aspects of everyday life can be temporarily found and enjoyed.

Clubbing and the politics of the self

Dancing is a play form in which, through flow, a sensation of usually private, often covert, and very much personal vitality can be experienced, although clubbers will tell differing stories about the nature of this vitality just as they tell differing stories about their experiences of clubbing.

BEN: ...what goes on inside your head?

SEB: That's another thing that was different about that night. If I'm somewhere where I know...I know The Tunnel I suppose...how can I explain this...if I'm totally relaxed, it's like I was saying to you before I went out, then the death thought comes into it, the writing of letters in my head. I think I wasn't free to do that on Saturday only because I was conscious of looking for Sarah, partly because I was uncomf...well not uncomfortable but just because of the unknown between us on our first night out. Partly because I was drunk as well. My mind kept flitting away to things like – 'Is Ben having a good time?', 'What else can I do to show him what I do in a club?' 'Is Sarah going to arrive?' 'Is that her over there?' – there was too much going on inside my head for me to totally relax and I don't remember relaxing really. I mean, how I want people to remember me, I am at that point the most complete person I could be, the happiest person that I could be, I'm just...you know, everything has a purpose and everything has meaning, and nothing has meaning and nothing has a purpose – you're just there on your own and everything: brilliant! Ummm…but if it was all to end the next day as a result of drugs, I would only want people to think good things and to know that it was a choice that I made and I stand by that choice and I'd want them to know that I had a most fulfilling time ever. Ummm…I think about writing something that I'd like to be read at the funeral, by someone who knows me. I'd think of the tunes that I'd want to be played. I would want to go out to Grace, 'It's not over yet' [a popular anthemic track] and another tune, a banging-hard-happy tune. I would want people to go to the funeral in their party gear – none of this black tie sombre stuff. Go the way that they remember me. And if people started dancing in the aisles of the funeral parlour then all the better. Better to be a celebration of life and not a sorrowful experience or a good-bye.

In this starkly melancholic and reflective story of Seb we begin to gain an understanding of quite how important clubbing is in Seb's life. Seb explains how, at certain times during his clubbing experiences when he 'is totally relaxed', his thoughts turn to in some way capturing that moment as 'his essence'. Clubbing is what Seb *is*, it is what he is about. I am wary of reading 'behind the lines' of this extract too much for I have no idea what other aspects of Seb's life may be triggering these thoughts, but I do want to draw out just two points. First, Seb contrasts being 'totally relaxed' on one hand, with the particular night he is talking about on the other. During the latter, where he was worried about me, about his girlfriend and about being drunk, these worries prevented him from the deep contemplation or 'flow' of which his 'death thoughts' were part. This frustrated him as it was not how he wanted 'to be remembered'. Second, Seb notes how, when he is relaxed, he is 'the most complete person' and 'the happiest person' he could be. '[E]verything has a purpose…and nothing has a purpose' perfectly captures the strong sense of self-validation that may characterise playful vitality. Of course, other clubbers will experience and evoke the sensation of playful vitality in different ways.

BEN: What do you think about when you dance?

MARIA: Errrr…I suppose fucking stumbling all over the place not really being…sort of bumping into people, sort of looking around to see how other people are enjoying themselves, ummm…if I'm dancing a lot and I'm enjoying myself I might not be thinking of that much. I maybe think about how the evening has sort of developed and how I'm feeling now [right] ummm…yeah.

BEN: Does your mind ever wander about things outside that particular night?

MARIA: Yeah, all the time, when I'm feeling...when I'm dancing...when I'm having a good time I forget about the worries that I have, like what the fuck am I doing in my life, you know, um...what am I going to do about that? I think about these things but I think 'don't worry, you're only 24, um...you've got a lot of time to sort things out, something will sort itself out, you know. You're intelligent, you've got a lot to offer, you know, don't worry about it'.

Here, Maria is expressing broadly similar sentiments to those of Seb (p. 152), although in a less explicitly melancholic fashion. Whilst part of her consciousness is occupied with the spatial and social demands of the dance floor – '...looking around to see how other people are enjoying themselves' – another part of her consciousness is concerned with wider and deeper issues – 'what the fuck am I doing in my life?'. The time–space of dancing is for Maria partly one of contemplation, affirmation, self-validation and resolution. Again, this perfectly captures the rewarding and reflective nature of the experience of playful vitality.

In these extracts, Seb and Maria are not resisting any notion of 'authority'. Rather, they are engaging reflexively with *themselves* and aspects of *their* own lives. Their vitality is drawn not from a political engagement with 'others' or a dominating, subordinating exterior force, but from an inward exploration of *this* particular aspect of their identity, *this* identification. In the case of dancing then, playful vitality is a form of oppositional practice in which the primary scale of operation is the bodies and experiences of the individual dancers, and one set of foci for these oppositional practices is alternate facets or identifications within the clubbers' own lives.

Clubbing and the unshackling of individuality

Instead of yielding to a music that is more or less identity-affirming, the sensual activity of losing oneself and regaining something else on the dance floor opens up a social space that is quite different from the public and private boundaries that hold our identities all too tightly in place.

(Ross, 1994: 11)

SEB'S JOURNAL ENTRY: If anyone can claim to have had a better time in a better place (with or without the aid of drugs) than we *all* have when we're together on Saturday night at Supreme, then I think they're either lucky (*mild* understatement) or lying (much more likely!). I know I've said it before and I know that you all realise it too, but what's wrong with saying it again? *We're all wonderful!*[63]

The communality of the experience for Seb is once again apparent here, but it is the powerful sense that this communality has engendered Seb with a deep sense of *personal* worth and vitality that marks out this extract from Seb's clubbing journals as interesting. When he writes 'We're all wonderful!', Seb is not being arrogant or cocky – this is for 'internal distribution' to only those who were there – rather, this is an expression of what he and his friends already know, and is an indication of what clubbing with his friends does for Seb – it simply makes him feel great.

The clubbing experience not only provides opportunities for a reflexive exploration of the self, then, but is also very much a social encounter. Just being within the dancing crowd can be enough to instil feelings of vitality within an individual (Le Bon, 1930). Thus, playful vitality is constituted both *individually*, and in one sense privately, but also *communally*, and thus socially. Although being together is a basic given in society (Maffesoli, 1995) – the social is what makes society – certain contexts of being together in social situations are more pleasurable than others. Being together with many others in the clubbing crowd can reinforce the vitality experienced through the experiencing of flow on an individual scale, with the crush of people together magnifying the sensory stimulation of the mediations which constitute a critically important spacing of the clubbing environment: lighting, warmth, loud music, continual movement, semi-visibility, verbal obliteration.

In the second stage of 'The night out' I introduced Bauman's notion of 'manifest togetherness' in unpacking one understanding of the clubbing crowd (1995b: 47). In certain instances, this 'manifest togetherness' can 'numb through overstimulation: ecstasy leads to (borders on, melts with) nirvana' (ibid.: 47). Bauman adds, 'togetherness of this kind is mostly about the "unburdening of individuality"' (ibid.: 47).

> Ostensibly, the masks are torn off to uncover the grimace of a bare fact; in fact, it is faces that are wiped clean of their identities so that another identity may rule supreme which is no one's identity, no one's responsibility and no one's task.
>
> (Bauman, 1995b: 47)

This collective unshackling of identity that can characterise certain forms of group encounter is a further important strand of playful identity, for it is this betweeness of self and crowd – the flux experienced between experiencing oneself egocentrically and experiencing oneself lococentrically whilst dancing – that facilitates both moments of individual reflection *and* moments of ecstasy (or *exstasis*), at times simultaneously. This subsumation of the clubber into the clubbing crowd and yet also into him or herself is reminiscent of Rossi Braidotti's

(1994) work on the 'nomadic subject'. For Braidotti, 'nomadic consciousness consists in not taking any kind of identity as permanent' (ibid.: 33; cited in Pini, 1997a: 127), and Pini (1997a) proposes that this nomadic metaphor neatly evokes the situation whereby a 'raver moves into the space of an event, makes connections with other bodies...and leaves, never likely to meet any of these bodies again' (ibid.: 127).

Clubbing, difference and sentiments of inclusivity

Forms of collective effervescence (Durkheim, 1912) such as those experienced through music and dancing have often been conceptualised through the employment of Victor Turner's (1969; 1974) concept of *communitas* (see Spencer [1985] and O'Connor [1997]). In brief, proposing that political power is fragile, and that all humans are united by a shared experience of the mysteries of life, death, fertility and the uncertainties of nature, Turner suggests that humans are essentially humble beings who live and thrive only because of their involvement with, and interaction between, each other in the form of communities.

Spencer argues that the associations between dancing and *communitas* 'should be immediately apparent' (Spencer, 1985: 28), with O'Connor paraphrasing Turner in noting how one of the major ways in which a sense of community is achieved today is 'through the breaking-down of hierarchical structures which form part of everyday life, the abolition of difference and the creation of a sense of "communion"' (1997: 154). For O'Connor, the *communitas* of the 'set-dance' (her object of study) 'consists of people who may have come from different social backgrounds but who are rendered equal by the nature of the set-dancing encounter' (ibid.: 155).

O'Connor's account of *communitas* on the dance floor over-simplifies and idealises the encounters that take place there. Whilst it is certainly the case that people who may not know each other dance together, as O'Connor proposes, it is wrong to assume from this that the dancing crowd consists of socially very different individuals who generate 'instant community' (O'Connor, 1997: 155) through dancing. Throwing off the shackles of identity does not necessarily mean that those identities were that different to begin with. Furthermore, to take clubbing, this would also be to gloss over the centrality of notions of style and display, of coolness, of 'appropriate behaviour' and its ad libbings, and of skills and competencies, all of which are used to differentiate – as well as establish identifications with – other clubbers within the dancing crowd.

In his somewhat rambling discussions of communality and difference, Maffesoli (1995) attempts to build upon Turner's (1969) notion of *communitas* and introduces the concept of 'unicity' in proposing that the sharing of a space and the social situation within it can act to bind that social situation through the generation and consolidation of a group ethos. While this crowd 'may be made up of a plurality of elements...there is always a specific ambience uniting them all' (Maffesoli, 1995: 14). Maffesoli loosely defines 'unicity' as a state of togetherness that goes beyond the (rationalist) 'dream of unity', instead being characterised by the 'adjustment of diverse elements'. There is no magical increase in tolerance of difference, but rather an 'organicity of opposites' (ibid.: 105) in which the notion of 'outsider' or 'stranger' becomes transferred onto those *beyond* the social situation at that moment, in that place. The 'mutual tuning-in relationship' (Schutz, 1964: 161) that characterises the interaction of diverse elements in this way is, suggests Schutz, the bedrock of all communication.

While the 'dynamic rootedness' that for Maffesoli (1995: 107) characterises his notion of unicity is premised upon the heterogeneity of crowds being energised through exposure to diversity, Maffesoli's conception suggests an all-too-easy, somewhat inevitable, empirically unproven and almost miraculous thriving on differences when it is quite clear that this is not an inevitable outcome. Admittedly, dancing can be seen as central to the disrupting of cultural, gender, age and ethnic boundaries (Spencer, 1985), with an emphasis on the momentary and the now, but this does not mean that 'anything goes' in the clubbing crowd. Extensive cross-genre research into the attitudes and practices constituting and differentiating each of the many genres of contemporary clubbing, as well as detailed studies of the actual constitution of clubbing crowds, would be necessary for one to draw any firm conclusions about if, when and how clubbing crowds in any way dissolve or obscure social differences or if they are, in fact, comprised of more varied identities than any other form of play. For now, the experiences of the clubber interviewees blended with some theoretical contributions concerning similar social situations must suffice.

To start with, clubbing is no utopia — a concept of an imaginary place (literally a 'non-place') that by definition can never be realised. Rather, through clubbing, its moments of ecstasy and the playful vitality that can be experienced, a type of *utopian sentiment* may be achieved or inhabited. Richard Dyer (1993) usefully makes the critical distinction between *models of utopian worlds*, such as those of Wells (1906; 1933) and More (1965), and the *feelings of utopianism* that may be experienced through, and form the *raison d'être* of, many forms of contemporary entertainment. The qualities of experiencing 'wish-fulfilment', 'escape', 'something better' (Dyer, 1993: 273), a 'dream world' (Spencer, 1985: 34), the 'imagination of nirvana' (Storr, 1992: 98) — these are all utopian feelings *as experienced* rather than the parameters of a utopian world as, for example, More and Wells set out in such detail (I discuss Dyer's notion in further detail on p. 160).

Writers in the 1960s and 1970s such as Huxley (1961; 1972) and Leary (1968; 1973) tried to convince young people, and indeed, society as a whole, that utopia was already here within their grasp in the form of psychotropic substances.

Yet, despite the use of similar substances in some clubbing experiences, the clubbers appear aware that the worlds they inhabit through the practices and spacings of the dance floor are imaginary and temporary, yet no less meaningful for being so (Dyer, 1993). As the following extract from the first interview with Kim and Valerie suggests, clubbers can and do idealise their experiences.

BEN: So what is so good about it?

KIM: The whole atmosphere...

VALERIE: ...it's electric...

KIM: ...everybody's happy and just there to enjoy themselves...

VALERIE: ...everybody's up for it and they're just loving it and it's brilliant [uh-huh?].

KIM: The music used to be really really good there and it used to build really really well [uh-huh] and it used to take everybody sort of through the night on the same level and you'd be like 'wa-heyyyy' and like smiling and chatting to people that you've never met before or just the occasional nod on the dance-floor. I suppose Banana Split because that's where I found that sort of club the first time that I went there and it was like 'wow' [yeah] and I don't suppose there are...although they are a lot of clubs – they're all going to be slightly different. We were just lucky to find one like that. We've been to a couple of others, like Fierce Child, and that and they were good but...

VALERIE: ...we went to Elysium didn't we...

KIM: Yeah, and that was good, but that was way back in the early days and...

VALERIE: Yeah, and we didn't really know what we were looking for, yeah [uhm-hum]. We just went looking for a night out and the nights were just...to this day I haven't seen nights like that in my life, and it was really a good club. I find that gay clubs tend to be a lot more open and everybody's up for it. Banana Split's supposed to be a gay night isn't it?

KIM: No, it's not, it's just...I think it's just 'open' as well.

VALERIE: Right, but Elysium was good although we didn't really know what we were there for – we were just there to go out [right].

KIM: And then you've got the music – we didn't really know what sort of music or...

VALERIE: But you're not on your own though, ever. Someone's always smiling at you, talking to you, dancing at you, or whatever.

KIM: It's like you're at a party and you know everybody there, it's wicked.

BEN: What are the other clubbers like?

VALERIE: All sorts – the only thing that they have got in common is that they go out clubbing, transvestites, girls in their little skirts and their spangly tops, people in trendy trendy club stuff, there's like people in workman's jackets and you've got all the lads in there in like dealer jackets, black men in their sunglasses. There's people that look twelve in there, yeah. I've seen a couple of really young people in there the last couple of times that I've been.

KIM: That's what's so good about Banana Split – it's not a trendy club, it's not a hard club, it's not a trance club – it's just lots of people having fun, that's all.

From this captivating exchange between Kim and Valerie I want to draw two points related to the present discussion. First, they spend the initial part of the exchange saying how incredible the club, Banana Split, is – 'I haven't seen nights like that in my life...'. The music and the people stand out as particularly important. Second, at my prompting, they then go on to discuss the constitution of the clubbing crowd at Banana Split, fleetingly surmising that it is a gay club because of its apparent 'openness', although it is not (and Kim points this out to Valerie). Kim and Valerie then list the 'different types' of clubber that go to the club, saying that 'the only thing that they have got in common is that they go out clubbing'. My impression of Banana Split, gleaned over four visits, was that the range of differing identities there was actually extremely narrow. My research diary for one night I spent there records at one point:

RESEARCH DIARY EXTRACT:[T]he clubbers were mostly between 17–20 years old, with very, very few people over 23, I'd say. Probably slightly more men than women; almost zero range in ethnicity – practically all white...as far as style was concerned, this was interesting: for the boys, jeans and either trainers or boots (lots of Ralph Lauren in evidence), short cropped hair; for the girls it seemed important to wear as little as possible, frequently little more than just underwear, and lots of boob tubes and short skirts, lots of fake tans, chests crammed into Wonderbras. Overall, there seemed very little variation in style, dress, identity...most were HAPPY, that's without the slightest doubt, but most were similar to each other, at least they seemed that way...

Building on this disjuncture in perceptions in attempting to deepen understandings of how different identities constitute clubbing (or not), Goffman's (1961) shrewd analysis of social encounters can, once again, be of assistance – in this case, through his work on the role and nature of differences within a social gathering. In his discussions of the sociability of parties, Goffman suggests a party is: 'by way of being a status blood bath, a leveling up and leveling down of all present, a mutual contamination and sacralization' (1961: 75). However, Goffman qualifies this by noting how, in his analyses, those invited to parties were actually more similar than they might have imagined themselves to be, but that, at the same time, a 'feeling of boredom, that nothing is going to happen' (ibid.: 75) can occur when social encounters are entirely made up of those that know each other or are socially similar.

> So we find that the euphoria function for a sociable occasion resides somewhere between little social difference and much social difference. A dissolution of some externally based social distance must be achieved, a penetration of ego-boundaries, but not to an extent that renders the participants fearful, threatened, or self-consciously concerned with what is happening socially. Too much potential loss and gain must be guarded against, as well as too little.
>
> (Goffman, 1961: 75)

It seems that clubbers, in a similar fashion to Goffman's party-goers, might assume greater diversity within the dancing crowd than there actually may be. The notion that the dancing crowd can act as a source of group effervescence *because* of this supposed diversity, as alluded to in discussions of *communitas*, unicity and utopia in points made by O'Connor (1997), Spencer (1985) and Maffesoli (1995), is not entirely convincing. This is not to deny for a moment the validity and strength of the sentiments of inclusivity of which some clubbers talk. These clubbers do, it appears, experience sensations of openness, both resulting from the use of empathogens such as ecstasy (MDMA) and as an outcome of the general euphoria and elation that can characterise crowd-based oceanic experiences. However, rather than suggesting that because these sentiments were 'real' – they were experienced as such by some of the clubbers – then the clubbing crowds miraculously became the theoretically slack, yet poetically evocative 'melting pots' of different identities that often cause one to cringe in certain accounts of the city, it might be more useful to return to the distinction between 'sensibilities' and 'reality' that Dyer (1993) makes in his discussion of utopia and entertainment.

Differentiating between a sense of 'utopianism' on one hand and 'models of utopian worlds, as in the classic utopias of Sir Thomas More and William Morris' on the other hand, Dyer suggests that the utopianism present in various forms of entertainment – and his 'case' is the Hollywood musical – is 'contained in the feelings it embodies' (1993: 273). Dyer proposes that the Hollywood musical industry expresses the *needs* of those that come to enjoy the musicals – and especially their desire for a better and a different social order(ing) – firstly through 'non-representational means' (ibid.: 273) – music, colour, movement – and secondly through presenting social relations between people more simply and directly than they exist in other aspects of lived experience.

Thus, in the musicals, the stars always seem 'nicer than we are, characters more straightforward than people we know, situations more soluble than we encounter', yet, at the same time, the nature of the non-representational 'signs' that 'we are much less used to talking about' are also important – the beauty of the melody, the movement of actors on the stage, the expressions on their faces, the speed and excitement of the performance, the smarter-than-usual dress of those co-present (ibid.: 273). Together, these qualities distil the utopian sensibility that is experienced as 'real' by the audience for the duration of the performance. The feelings that these qualities elicit from the audience are, in one sense, what it would feel like to inhabit a 'real' world like that on the stage.

In a parallel fashion, we can begin to understand how the clearly powerful expression of playful vitality, which it is possible to experience through clubbing, can instil utopian notions within the clubbers' imaginations and can be based upon idealistic visions of the way they would like things to be (or know things not to be) in other aspects of their lives. Dyer sees utopian 'sensibilit[ies] as temporary answers to the inadequacies of the society which is being escaped from through entertainment' (1993: 277). In a similar way we can understand the notion of vitality through play and the apparent openness and inclusivity on the dance floor as an ephemeral relief from a society in which communality and trust are regularly swamped by individualism, communication with 'strangers' is seen as an inherently and increasingly 'risky' activity, work dominates play and oppressive (as opposed to expressive) codes of social interaction seemingly cut through virtually every social space through which we live our lives. Lastly, what makes clubbing an even more powerful site for utopian sensibilities than the Hollywood musicals of Dyer's analysis is that, in clubbing, the audience are also the actors – their utopian 'worlds' are of their own making[64].

'It makes you more free'[65]

Playful vitality is only one conception of the sense of inner resolve, pleasure, resilience and bodily control that can be experienced through the combination of very loud music, dancing to that music within a dancing crowd, and the spatial and visual effects (the mediations) through which clubbing is produced as a sensory experience. The sensation of playful vitality is problematic to evoke for it is rarely articulated through direct speech by clubbers. For many it is the subconscious reason why they enjoy clubbing. For others who are more conscious of its value, attempts to describe and evoke the sensation, even for the most articulate of clubbers, often falter as they confront the limits of spoken language.

BEN: What does dancing 'do' for you then?
BRUCE: Dancing is your body's expression of that feeling – there's ummm...this
    is going to sound appalling, I'm sorry [we laugh] – there's a book by someone
    called Robert Liston called the *Salamander Tree* and it has a very sort of

complex plot, but it's set in the Second World War and it's all about an alchemist, umm [yeah] who...the real end of alchemy is not to turn lead into gold, the real end to alchemy is to find the sort of key to life and the spirit that inhabits us all and ummm, in this book an alchemist actually finds this, and once he's got it he wants to find a form in which to trap it and so what he does is he paints it and turns it into a paint, and paints it and that actually takes me completely off...[we laugh]. Well, hang on...right...in this book, there's this plot and the plot goes that umm...during the war the Nazis were doing this experiment – this is going to sound so fucking bad – in which they monitored the electric signals that were given off by people just day to day...

BEN: [incredulously] Is this part of the story true?

BRUCE: I think it might have some sort of basis...they just monitored the way in which human beings affect the atmosphere around them electrically and then they turned those electric signals into music, they sort of plotted them on a diagram and then they turned them into music, and they played that music back to people and they found that that music made them dance and that that dancing in turn gave off another set of signals [right] that was sort of the key to life, if you see what I mean...

BEN: [slightly sarcastically] I know exactly what you're saying...

BRUCE: And then that was turned into paint...anyway...it's really a very good book!! [yeah]...so, dancing is doing that in a sense, it's releasing the way in which everything reacts around you through your body, and when you have house music played to you it sometimes taps into that, and in combination with drugs as well it makes you more free and able to do that and that's just what you do, and it's tied in with sex and it's tied in with looking and the seething and so many people sort of mimicking people on the dance floor – you know you'll see someone doing a particularly good move and then someone next to you or you will try and do that move [yeah], and you can either do it better or you realise that it's an absolutely stupid thing to try and do with your body and...So, it's all to do with that, and it's to do with sex as well – you see women dancing and you think 'bloody hell if I dance then they'll love me' and so it's all that knotted up together.

BEN: How does that music then work at home, in the car, on the Walkman?

BRUCE: Because it's redolent and it reminds you of those times when you have been out dancing to it. I think, I mean it would be very hard to get someone into house music who hadn't been clubbing because it's an experience that's first-hand, if you see what I mean, and what it does to you [ummm] so when you hear it at home it has its value and its ummmm...worth as music, and because music will always affect you and if it's good music it will always affect you like good music [yeah], but it also carries with it memories and you know sometimes expectations – I've had the experience where I've

heard a really good song on the radio and thought I'd fucking love to hear
that in a club because I know that on a good sound system and stuff it would
do a lot more to me than if it was just there on the radio, but on the radio it
sounds good anyway, so ummm...

As was the case with many other clubber interviewees, the feeling that Bruce
actively thought about the nature and 'purpose' of his clubbing experiences on an
everyday basis, and often in some depth, is communicated clearly in this short
extract from the first interview with him. As an older clubber (aged 26), at least
in terms of those that I talked to, the comparison of the stage at which he is now
compared to a younger 'teenage' mentality is perhaps predictable, yet still significant. Within the first paragraph he has made the connection between music and
dancing, suggesting that the most pleasurable way to listen to (in his case) 'house'
music is in fact to dance to it, for dancing expresses feelings and sentiments that
are perhaps far more difficult, and even at times impossible, to express through a
concentration upon verbal means alone. Some of the subtle techniques of the
dance floor are touched upon here by Bruce: the senses of looking, of mimicry, of
seething crowds, of sexual interaction or, at least, sexual display. The lengthy
anecdote about the *Salamander Tree* merely reinforces the sense of constant evaluation and even philosophical investigation that goes into Bruce's clubbing
experiences both during and after the night out. In many ways, the simplicity of
his story, re-presented here verbatim, evokes more sensuously, forcefully and
poetically the nature and importance of clubbing for Bruce than any structural
analysis of dance or Deleuzian-inspired notion of 'musical molecules' or
'threshold of disappearance' (Hemment, 1997) despite the seductive and
provocative nature of these latter descriptions. Finally – and most powerfully of
all – throughout the extract there is a palpable sense of the vitality that Bruce
draws from his clubbing experiences.

GOING WITH THE FLOW

This fourth section of 'The night out' has been about developing an understanding
of the experiencing of playful vitality – an experience that, while tricky to pin
down and certainly problematic to articulate verbally, appears to be centrally
placed in the clubbing experience. The conception of playful vitality suggests that
conventional accounts of both play and resistance are inadequate in understandings of contemporary social formations. However, whilst I would agree with Gore
(1997) in her suggestion that the description of rave as 'a ritual of resistance'
(Rietveld, 1993: 41) is a misnomer, I would also argue that her alternative
conception, premised on the notion that this 'ritual of resistance' approach

ignores 'the explicitly apolitical stance of many of its participants' (Gore, 1997: 51), is also flawed. The flaw lies primarily in the simplistic construction of the political, a construction in which the political is inextricably linked to narrow preconceptions of the relationships between power and resistance, and between 'authority' and subordination.

The vitality that can be experienced through dancing takes the form of a sense of individual and communal euphoria, induced through the playful practices that constitute dancing, as well as the specific contextual details that mark clubbing out as different from, say, line dancing, ballroom dancing or even disco (but that's another story altogether!). The vitality experienced through dancing during clubbing is largely emotional in constitution, arises partly through the flux between self and the dancing crowd, and prioritises atmosphere, the affectual, proximity and tactility, and the here and now. The playful vitality of clubbing is communal, but unlike the communities ordained by politics, economics and histories – in short, by rationalism – the communalities of the dance floor are constituted by the instant and the ephemeral, by sociality, empathy and non-rationality.

Playful vitality, then, is not 'resistance' as conventionally theorised, especially with reference to so-called 'youth cultures' and their supposedly resistant practices – this is not a resistance to a parent or 'dominant' culture. Playful vitality is, rather, partly a celebration of the energy and euphoria that can be generated through being together, playing together and experiencing 'others' together. Yet playful vitality is also partly an escape attempt, a temporary relief from other facets and identifications of an individual clubber's own life – their work, their past, their future, their worries. Playful vitality is found within a temporary world of the clubber's own construction in which the everyday is disrupted, the mundane is forgotten and the ecstatic becomes possible.

> [W]e have slowly come to appreciate that the realisation of resistance is not to be located in a prescribed future. It is already inscribed in the languages of appearances, on the multiple surfaces of the present-day world. There, re-working the relationships of power, resistance hides and dwells in the rites of religion, in the particular inflection of a musical cadence...it is where the familiar, the taken for granted, is turned around, acquires an unsuspected twist, and, in becoming temporarily unfamiliar, produces an unexpected, sometimes magical space...[I]t is our dwelling in this mutable space, inhabiting its languages, cultivating and building them and thereby transforming them into particular places, that engenders our very sense of existence and discloses its possibilities.
>
> (Chambers, 1994: 16)

Higher than the sun

My brightest star's my inner link,
let it guide me,
experience so innocent beats inside me,
hallucinogens can open me or untie me,
I'm drifting in a space, free of time –
I've found a higher state of grace, in my mind.

I have glimpsed, I have tasted,
fantastical places,
my soul, an oasis – higher than the sun.
(Primal Scream, 1991, 'Higher Than The Sun')

Everything's entirely completely wonderfully positive!
(Kim)

Part Three

REFLECTIONS

Helen pushed her way out during the applause. She desired to be
alone. The music had summed up to her all that had happened or
could happen in her career. She read it as a tangible statement,
which could never be superseded. The notes meant this and that to
her, and they could have no other meaning, and life could have no
other meaning. She pushed right out of the building, and walked
slowly down the outside staircase, breathing the autumnal air, and
then she strolled home.

(E.M. Forster, 1946: 26)

INTRODUCTION

Although we have left the various clubs, the night is not quite over. An important part of the clubbing experience is the period immediately following the clubbers leaving the club. This is often a time of physical and emotional exhaustion, yet this period of 'afterglow' can also be a time of reflections.

This final part of the night – experienced as neither night out proper nor as quite yet back to 'reality' – is split into three stages. First, we remain with the clubbers as some of them stay together, and I present three stories of afterglow in which clubbers reflect upon their clubbing experiences. This period of reflection is central in consolidating and partly rationalising the clubbing experiences that have just passed. For some clubbers these reflections will be focused around re-living the night, while others will look forward to the days, weeks and even years to come. Still others attempt to make sense of the night out in the context of their everyday lives more broadly.

Second, in the light of the night out and in some ways following the clubbers' reflections on their experiences, I re-address and reflect upon some of the issues raised in the three broad starting points that I sketched out in 'The beginnings'. I make three points – about young people and resistance, about experiential consuming and communality, and about imaginations and betweeness, and I make these points explicitly through the lens of 'The night out' that has just passed.

Third, I conclude by noting how clubbing, through its particular fusion of motion and emotion, is important in many different ways to those that enjoy its often alternate practices and spacings. While at best only beginning to evoke the intensity and complexity of clubbing nights out, I claim that this book does represent an addition to our developing understandings of the practices and spacings that constitute the experience of clubbing.

THREE STORIES OF
AFTERGLOW[1]

The immediate post-clubbing period of the night out is frequently one of reflective recoveries – whether in small groups or alone. While clubbers will have had
unique experiences, it can be fun to establish common threads that have run
through 'their' nights. This can act to prolong the night and somehow defy the
temporal restrictions of licensing, the spatial restrictions of the venue closing, and
the general march of time towards the notion of 'morning', even though dawn
itself may have passed some hours previously. Other clubbers prefer to talk about
and reflect upon the night, in some way beginning to order the chaos of which
they have been a part just a short time previously. There is often the need for a
place to spend a few hours chilling out, drinking coffee and maybe eating, or
smoking some cannabis. Although not part of the clubbing experience in the
narrow sense, in that they do not usually occur within the club, these processes
and practices of reflection and the attempts at interpretation and understanding
that can occur in the immediate post-clubbing come-down, afterglow or 'chill-
out' period perform an important role for the clubbers. This stage of the night
thus represents a vital transitional phase between the night out and the wider lives
and other identifications of the clubbers[2].

Some of the post-clubbing activities and reflections of three clubbers or groups
of clubbers have been selected in outlining the important, yet varied nature of
this period of 'afterglow'. While each clearly differs, similar sentiments and attitudes – towards the night just gone, towards the week or days ahead, towards
each other and towards themselves – are recurring features. Three of the themes
that arise in slightly varying form throughout the following stories are: the notion
of a 'world of clubbing' opposed to a 'normal' or 'straight world'; the notion of
the spaces and spacings of clubbing opposed to the spaces and spacings of other
forms of public spaces; and the contrasting of clubbing with other facets of the
clubbers' lives. I return to these themes after the stories.

Kim, Valerie and John – holding on to the night?

Kim, Valerie and John live in outer London, on the border of Kent. For them, the post-clubbing period is about reflection, recovery, sharing their experiences and engaging in quite different forms of interaction to those of clubbing. This period is also dominated by travel and the impulse to get home – to find comfort – which is a forty minute train journey away.

BEN: What happens after you leave the club?

JOHN: Uh…things like going to other clubs…I mean I've only been to Tempo twice, three times now – that's Gray's Inn Road. I don't know, you know that you can carry it on afterwards – or you can go out Sunday night or you can go out – you can go out any time you like. It's not just stretching it out, it's the fact that you can take your pick – like in Brighton, you want to go out late you go to the Paz, you go to London, you can go to fucking anywhere – it's going to be open until four or five or six, there's always going to be somewhere to go. There's going to be a lot of people about going to these places [right] as well. Like walking through Leicester Square at three o'clock in the morning there's going to be millions of people everywhere. It blows me away that does. I…I find it…I was going to say exciting, it…it's the fact…it's dark, it's late at night, it's capturing the time that everyone else is asleep or…[yeah]…It's almost during the day, y'know, you can walk through Trafalgar Square, or whatever, and you're just another person, but at night, there's people everywhere like you and it's YOU'RE place, your time.

It is clear that the quality of the night out as a special or extraordinary experience does not abruptly finish with John's exit from the club. For John, and for Kim and Valerie, with whom he often goes out, clubbing is very much about *travelling* to and from the club as well as his time actually in the club. Having written off the local 'small-town' clubs these three and other friends of theirs from their town travel regularly to clubs in Brighton and in central London. The relative absence of the restrictions of club licensing hours in London is something that John values, partly because it means that he does not *have* to go home unless he wants to. This notion of John being in charge of himself – of the night-time spaces of London having a special aura of *freedom and alterity* that the day-time spaces seem to lack – comes through strongly towards the end of the extract. The inherent vitality of clubbing – the blend of dancing, music, crowds and friends – is experienced through a different lens once John leaves the club. Late-night London appears to 'belong' to him – it is his place and his time – and he decides what he wants to do, where, how and with whom.

Perhaps for the first time since they went into the club many hours previously, the group are now all together at once and can talk to each other easily in a group context. Thus, in addition to taking place against the backdrop of travelling, this post-clubbing period is also one of catching up and finding out about others' nights. It is a period of swapping stories of musical and perhaps chemical highs and lows, of how the night just gone compares to others in the past, of who they have met, what they thought of the DJs and the atmosphere, of how they feel right now, what they want to do now and where they want to go. The night is by no means over as a 'night' for Kim and Valerie (see p. 173), although much of what remains of it will be set in one of their sitting rooms. For Kim and Valerie this continuation is inevitable because, at least on the occasions on which I accompanied them, they will have taken ecstasy (MDMA). Sleep would be difficult if not impossible for perhaps another eight or ten hours, regardless of how much energy they might have expended on the dance floor. While they are no longer experiencing the ecstatic states that they talked about earlier, the effects of the drug upon them persist. Their perceptions of – and thus their relationships to – their surroundings and their experiences of themselves and of others, remain altered. Interactions between them are marked by a continuing openness and a closeness which, although obviously based partly within their pre-existing friendship, is magnified through both the shared experience of the night out and the continuing empathic effects of the drug ecstasy (MDMA) on their 'self-defences'.

Marghanita Laski is careful to point out how an ecstatic state will often extend *beyond* the 'tumescence and discharge' of the ecstasy itself, suggesting that such a state 'does often occur after an ecstatic moment, lasting up to an hour or so, sometimes more, sometimes less, and in this period the ecstatic moment may be realised, appreciated, interpreted, while the ecstatic recovers from the shock of the ecstatic moment' (1961: 58). It is during this period, adds Laski, that the person comes to appreciate 'what it was he [sic] believed he had felt during the sense-disordered moment of ecstasy' (ibid.: 60). It is not in the 'heat of the moment' that a careful consideration and interpretation of the clubber's experience is attempted (or is possible), but rather in the post-clubbing period of reflection and come-down.

It should again be emphasized that by no means all clubbing experiences involve a period of afterglow of precisely this nature. For example, most of the people in Laski's study group 'barely interpreted [their experiences] at all' (1961: 62), although most of their experiences were by contrast experienced alone. Indeed, it bears repeating here that many clubbers, including those who have taken drugs such as ecstasy (MDMA) during the night out, will not have experienced any ecstatic or oceanic states whatsoever. In any case, and related to this first qualification, it should be noted that the oceanic and ecstatic experiences are not pre-conditions for reflection, but rather simply represent a specific route into its practices. Clubbers who have not had oceanic experiences may also engage in processes of reflection. In the cases of Kim, Valerie and John the 'togetherness' of the night, which acts as a foundation for these reflections, is usually carried forward deep into the following day.

KIM: Well, we all leave together, obviously, and get a train or taxi. Generally, the next day is spent together.

VALERIE: This is what happens, right. We leave Banana Split and we're all bollocksed, and we generally get a cab back to Dartford and you get a cab driver who will either do your head in or they will...[laugh]...

KIM: ...sometimes when we've got a train and we've got a cab from Dartford, that's really done our head in because the cab drivers in Dartford are just too straight – they haven't got a clue.

BEN: Whose place do you go to when you get back?

KIM: My place...usually drop another bit [of an E], usually a half or just a quarter, partly to help wind down or finish the night off a bit so it's not over too suddenly.

VALERIE: I prefer not to do any more when I get in and just to have a smoke.

KIM: ...I mean part of it is the staying awake and talking about it, and we'll sit and have a smoke and argue about who's having to make the tea [laugh]. We listen to lots of music. From there it'll kind of drift on...you've had an excellent night, you feel physically tired but in a nice way – you've had a good night.

VALERIE: When we sit there we talk about all the people that we've met, all the things, all the hallucinations we've seen, how many times we went to the toilet, that sort of thing!

I want to highlight just three points from this extract that are of interest in further developing an understanding of the processes of reflection which constitute an important final chapter in Kim and Valerie's night out. First, there is the notion that 'others just don't understand', that, for Kim and Valerie, what they have experienced is beyond description or translation unless you were there. This comes across strongly in their description of the taxi drivers who, it seems, 'don't understand' and 'haven't got a clue' because they are from Dartford and not from London (rather than because they are taxi drivers). Second, the night clearly continues after the club as a special experience. More drugs can be taken at home to stretch the night into the next day so 'it's not over too suddenly', to deny the responsibilities of daytime and the jobs and other routines that daytime represents. This further taking of small amounts of drugs – and especially the smoking of cannabis – may also aid the physiological processes of 'coming down' as the clubbers drift back into their lives outside of clubbing. Third, the post-club 'come down' is used to reflect about the complete range of events and minutiae that together constitute the night, from the smallest details (who went to the loo most) to the most personal and emotional accounts of, for example, hallucinations. Their thoughts are very much more on what has just been, as opposed to the week to come.

Antonio – collusion and reflection

This second story is extracted from the entry in my research diary for this night, spent with Antonio and two of his friends.

> We stumble out of the club at around 3-ish – Soho is packed with people, crowding the pavements and roads, looking and laughing – everyone appears happy. Some are in groups, bustling their way along noisily – others are alone, silent and walking purposefully, on their way…Cars crawl down narrow streets which are already impossibly full with cars, Vespas, people, thronging crowds. This wasn't 'late night' for Soho – the night had hardly started.
>
> Home was, stressed Antonio, completely out of the question – the post-clubbing coffee at Bar Espana is, I learn, an institution that had to be observed. Yet this isn't so much about the coffee itself, as where and the way in which the coffee is consumed. It's a 'scene-thing': it's about seeing and being seen, about looking at and being amongst the cool people who are having a great time and thus being cool yourself just by associating with them, just by being there. Okay, you can get a coffee around the corner for £1 and have a seat, but it's not cool. Here you may pay 50 per cent more, but you get 100 per cent more cachet. Bar Espana is about the public display of coolness, about stretching the night out, saying that tomorrow doesn't matter – tonight is all there is. It's the here and now of inside the club again, but this time outside of the club. The crowd, though relatively varied in constitution, are together, bound by caffeine and the cheesy Italian techno-pop that thumps through the crowded bar.
>
> Everywhere you look in Bar Espana, open twenty-four hours a day, 365 days a year, are mirrors, TVs and coffee. One of the places most ill-suited to accommodate large numbers of people that I have ever come across harbours one of the most hyped-up coffee-drinking crowds I have ever seen. It feels European, despite the distinctly un-Euro chill in the air, and, as my mind flicks back across the other clubbing come downs that I have endured in the course of this business, I feel lucky right now. I suppose I feel so good because I know that, while this isn't really me either – I don't wear this stuff usually, and I'm often tucked up in bed by

now – here I am laughing and even forgetting why I'm here. For a second, I don't feel out of place at all.

The four of us chat about the night just gone, laughing about the atmosphere which was more like that of a cocktail or cabaret bar than a club – the venue is an 'adult' revue bar during the day – about the bizarre Russian accordionist, Igor Oetkin, playing a version of 'Fernando', by Abba, and the immaculately dressed Count Indigo with his ridiculous cover version of the cult-60s classic 'Woman in Love', about the G&Ts at £3.50 a whack, about my huge tie, Tim's 'The Saint' suit, and the leggy transvestite drinks waiters/waitresses, about Jonny Del Rio performing a live cover of Neil Diamond's 'Money Talks'. It may not have been a dance club, but it managed to completely transport us to another time and space, although being cynical, after paying seven quid for that kind of kitsch I think one's imagination is primed for overtime. Prompted by the knowledge that I am supposed to be writing something on it and the notion that they thus want to help me the best they can – they say they're jealous but I don't believe them – us four chat about the night and come up with the idea that the whole thing was essentially a collusion, a staged event that relies on everyone for its success, from the people up on stage to us dressed for the occasion and willing to suspend our sense of disbelief for a few hours, and pay hotel prices for drinks. And I get thinking and connect that with what happens in other clubs, which are also about collusion and the suspension of disbelief, although the mode may be different, the stage less obvious, the distinction between producers and consumers more blurred, the styling less eccentric.

We sit around for a while, talking about how long we will go on coming out for these types of evenings, wondering how long we will belong here, and how long we will keep up this kind of performance. Are we getting too old?

Bruce – holding on...

Reflections can often be based upon and include recollections of previous nights out. These reflections may veer towards thinking about the future, about their place within clubbing cultures and upon how long they see themselves continuing to go clubbing in the way that they presently do. This latter point is especially evident with older or more experienced clubbers.

BEN: Well, how old are the others?

BRUCE: ...oh let's see, they're roughly sort of my age, in that they're 24, 25, 26 up to sort of mid-thirties, and when I e-mailed you something I mentioned this in that I think there is very much a whole generation of people who started clubbing when I did, and even before, who don't want to grow out of it and don't want it to stop. I was going out with someone who was 33 and she went clubbing and so did all of her friends you know and...so there are a lot of older groups going...I'm sure that a lot of people my age have now got to a stage like me where they have a regular thing that they go to and that they're happy with and their contemporaries go to as well, and it is subsequently an older scene and it will continue to go that way until we're like 50 and then God knows what will happen...

BEN: So how does the clubbing thing fit in with your wider life, especially what you're doing at the moment [trainee documentary film producer]?

BRUCE: A lot of it's in the open – obviously in my immediate peer group there's someone there who goes clubbing anyway and there's someone else who goes clubbing quite a bit, and so we talk about it quite a lot and that's an obvious, instant thing that we've got in common [yeah] and a sort of mutual subject that you can discuss. Umm, as for the people that I work for, I obviously wouldn't tell someone in a very senior position that I had a great E at the weekend, however...I'm working on something at the moment and the producer's about 29 and doesn't go clubbing, but was asking me about it before and I told him, and I don't feel I'd be in jeopardy because I've told him because he's of a similar age and knows that this goes on and can see that it doesn't rule my life. However, it has weaved its ways into my work in that I've made one film about clubbing – I would like to make more and you know I very much would like the two things to join up like that because I'm interested in clubbing more than just because I like clubbing. I'm not a DJ, I can't mix.

BEN: What's the drive behind the desire to make another film?

BRUCE: It's definitely driven by the lack of something there – if there were a weight of films, all about clubbing and they were very good then I would probably be looking for another subject. It's very much I think that I can possibly say something about clubbing that would...that hasn't been said before or could be said very well. At the same time I'm not a DJ and I don't particularly want to put club nights on, and I enjoy going to a club, and that's what I want to reflect. I wouldn't want to do clubs myself, but a lot of people that I know have done that and they've thought, 'I'm going to put a night on or I'm going to make music'. I'm not musical in the sense that I haven't got that side to me so...

BEN: Is this work any form of a quest for answers for yourself?

BRUCE: Not really because I think I know why I go clubbing and I know why I like it.

BEN: Do you feel then that you know what's happening in others' minds?

BRUCE: I think I know how my friends feel about it because I think that they feel very similar to me, and I know that some of them feel stronger about it because they have different links with it, but I know, yeah, I think I know the position that it holds within my peer group, definitely. And that's tied up again with this thing about it not being a political statement in the sense that we were saying. It's something that you do and you get on with [yeah]. If that's what it is then that's fine – there's no problem about that.

During this conversation with Bruce it became even clearer that his relationship with clubbing went further than just being 'something for the weekend'. For Bruce, clubbing had started to intersect with aspects of his work as a trainee film producer. Within this extract, there is a sense that clubbing is a form of identification for Bruce that extends beyond the physicality of the club itself. Not only is clubbing an experience through which he and his friends are electively and affectually bound together, but it is also a form of shared tie-in with others whom he does not know so well, such as his work-mate – clubbing is a 'mutual subject of discussion'.

As I suggested earlier, post-clubbing reflections and recoveries are often bound up with a broadly common set of concerns, thoughts and sensations which arise from the socio-spatial disjunctures experienced by the clubbers on leaving the clubs. First, in all three stories the clubber narrating compares and contrasts notions of the clubbing world from which they have just emerged with the 'straight' or 'normal' world into which they've stumbled. For Kim and Valerie the discrepancy of this experience is marked socially and spatially – the taxi driver in Dartford was 'just too straight' and did not 'have a clue'. In contrast, in my narration of the aftermath of my night with Antonio, I was struck instead by how far the suspension of disbelief – some of the codes, styles and atmosphere from *inside* the club – was prolonged spatially, temporally and socially into the city *outside* the club. This clubbing/straight contrast clearly becomes blurred when (for example, in the case of Kim and Valerie, and in a different way for Antonio and I) some of the spacings and practices of clubbing – loud music, drug use, crowds, night-time – were juxtaposed with some of the structurings of the world 'outside' clubbing. For Kim and Valerie this juxtaposition occurred in terms of their domestic space, for me it occurred in terms of the 'city'.

Second, clubbing seems to somehow constitute a qualitatively different form, experience or 'mode' of social space (privatised, enclosed, safe) from the world 'outside' (public, open, dangerous or 'other'). John is especially struck by this: 'it's your place and it's YOUR time'. Also struck, although in more subtle ways, is

Bruce who almost seems to reflect upon clubbing as a secret life – 'it's something that you do and you get on with it' – yet one through which empathy can be generated with others outside of clubbing.

A third theme encompasses both these previous points, and this is the tension between the clubbers' experience of clubbing and the clubbers' lives more broadly. Bruce's clubbing world is now important enough to be overflowing into his career, and yet he does not see this as a problem but rather as an opportunity: 'I think I can say something about clubbing'. For John, the anonymity of his every*day* experience of Trafalgar Square is contrasted with a different experience of this anonymity – not as individualising from, but as belonging to – that characterises his every*night* experiences of the Square. Finally, in my role as diarist for the night spent with Antonio and friends, I noted the tension between my clubbing persona and my researcher persona – a tension that the promptings and questions of those I was with served only to magnify.

Other lives, other identifications – clubbing in the city

Although my interest in this book has been upon the constitutive practices of clubbing experience themselves, the night out is clearly both of such critical importance to many young people and very much an urban phenomenon that it poses significant questions about our understandings of the social life of the contemporary city more generally. In particular, 'The night out' forces us to question 'which city', 'whose city' or 'whose cultures of the city' researchers of contemporary urban living choose to illuminate, for clubbing remains largely unstudied. As I have intimated throughout 'The night out', there are numerous reasons for this apparent absence of interest. Clubbing is of the night, of darkness, of the shadows. The music is often constructed as meaningless, lyric-less and even senseless. The dancing that is so central to clubbing is frequently represented as primitive, carnal and not worthy of serious study in any case. The moments of ecstasy that can, on occasion, characterise the clubbing experience and that appear to be able to engender such vitality within those that experience them, are seemingly even less worthy of study than the dancing practices of the clubbers. Furthermore, the social researcher is frequently ill-equipped (or too old) for forays into what is often perceived by 'adults' to be the 'ante room to hell' (Burke, 1941: v).

However, what Finnegan (1989) might refer to as these 'hidden cultures of clubbing' are constitutive of the city in the same way as the equally complex but somehow, for the researcher, more accessible, more respectable and more relevant city of working nine-to-five, of transport networks and environmental problems, of tourism and tradition, of architecture and planning, and of shopping and commerce. Paralleling de Certeau's (1984) contrasting of an 'ideal or plan of the city' versus an 'experience of walking in the city', future studies might re-place clubbing not only back into the other identifications and personae of the clubbers concerned, but especially back into the social life and spaces of the city. Together with the more formal understandings of city life there exist countless imaginations of the city which are imperceptibly fused to constitute what de Certeau calls the 'sieve-order' (1984: 107) of the city – the city of everyday and every night life. Deepening our understandings of the dark areas *between* the grids that form this 'sieve' would appear to be a potentially tricky yet conceptually crucial process, and one which I hope I have at least sign-posted through this study.

PLAYING – CONSUMING – FLUXING

The rather nebulous character of the conception which I have developed of how clubbing intersects with the wider lives of the clubbers and the city in which they live does not detract from the powerful, theoretical and conceptual resonance of the night out itself. Just as the clubbers engage in various practices of reflection about the night just passed, so I, in terms of this book and the many nights out which have formed part of it, want to reflect on where 'The night out' has taken us and how we might progress and deepen an evolving understanding of clubbing in terms of the starting points I set out in 'The beginnings'.

Clubbing matters to many young people. There is little stylistic or musical unity across the cultures of clubbing, and the already countless genres of clubbing continue to fragment and multiply. Yet for those that enjoy clubbing, it continues to constitute one source of vitality and an important form of self-expression within their wider lives, whatever the beats or other clubbers involved. Theorists and academics interested in clubbing cultures have begun to recognise this significance of clubbing for young people and – with varying degrees of success – have attempted to present understandings of how and why the practices and spacings of clubbing matter to the clubbers. Remarkably few attempts have managed to evoke the interactional complexity, the communality, the technical and stylistic demands and the emotional and imaginative constitution of clubbing. Ethnographies of clubbing are especially scarce.

It is no surprise that the experiences of the night out which we have just traced provoke many more questions than they provide answers – this is not something of which to be wary if one is searching for textured understandings as opposed to notions of 'truth'. However, in terms of our understandings of clubbing for eighteen clubbers in London during the mid-1990s, it is possible to make a number of connections with and progressions from the 'starting points' outlined in 'The beginnings', as well as drawing together a number of themes that subsequently cut through the four sections of 'The night out'.

Therefore, I now make three clearly related, overlapping, but as I present them

180

distinct sets of points which spin out of 'The night out'. First, I address issues relating to understandings of young people having fun. I propose that the playful vitality experienced through clubbing is no less significant to the lives of those who experience it than, for example, the so-called 'resistance' experienced and expressed by 'youth cultures' and student movements in the late 1960s. Second, I return explicitly to debates and questions about the relationships between identities, identifications and the practices and spacings of consuming as I further open out a conception of 'experiential consuming'. Third, I revisit discussions about betweeness and the understandings of diversity, which I began to critique in the previous section of 'The night out', highlighting the imaginative construction of clubbing as other-worldly.

Young people, fun and vitality

Energy is eternal delight!

(William Blake, 1969)

'Play' is implicated in the everyday lives of people in ways that have not been given adequate recognition in existing literatures (Thrift, 1997). Far from being a meaningless and peripheral activity which academics largely treat either as an 'optional extra' or – and even worse – as not mattering at all, play, in this case in the form of clubbing, can be seen to constitute an important context of and for sociality in the lives of many young people. For some of the clubbers to whom I talked and with whom I went out, clubbing was currently the most important aspect of their personality, impacting upon their work both directly (in the case of Bruce) and indirectly (in the case of Seb). Play, then, appears to be about far more than simple notions of 'pleasure' or 'recreation'. Play is also tied up with further and more complex notions of identity and identifications, and of vitality.

Second, previous understandings of young people and their so-called 'subcultures' have often conflated notions of shared appearance, and especially of similar attire or shared tastes in music, with those of identification and (thus) possible subversion. Theorisations premised upon the crude use of semiotics are particularly guilty in this respect (Hebdige, 1979). Many understandings of the ways in which young people have fun – in groups and individually – appear to prefer a distanciated understanding, from afar, from above, and usually from outside. Much less has been written about the detailed practices and spacings through which this playing is actually comprised. The practical techniques through which young people differentiate themselves from – and thus simultaneously affiliate themselves with – others, and from and with other groups during group encounters have been neglected or crudely over-simplified. These practical techniques and their spacings deserve further recognition and a more detailed understanding.

'The night out' infers the centrality to the clubbing experience of clubbers' understandings of the rhythms, beats and lyrics of music, and their expression of these understandings particularly through the use of their own bodies in dancing (Frith, 1996; Hanna, 1987). Not only does music act to intensify the emotional and imaginative impacts of being within the clubbing crowd in a manner proxemically and sensorially different to most, if not all, other facets of their everyday lives, but, partly through these processes of intensification and their alternate nature, the music can also act as a focus for an identification, almost in a totemic fashion (Hetherington, 1996; Maffesoli, 1995). It is thus less the music itself than the clubbers' responses to that music – their understandings of it – that are important. The complex, interactional spacings – the mediations, territorialisations, bodily practices and emotional resonances – that together constitute dancing add to the identificatory power that the clubbing crowd can exert upon individuals. The consuming of empathogenic 'dance drugs' such as ecstasy (MDMA) appears to further magnify these sensations of identification.

Through its privileging of music and dancing, clubbing demonstrates how young people's social identifications go way beyond simplistic notions of clothing and even style. Clubbing is constituted through a complex inter-weaving of continually unfolding practices, spacings and timings which, beyond a certain stage – and always less for some than for others – have less to do with distinction and the forging of notions of individuality and perhaps more to do with belongings and the establishment of identifications (Crossley, 1995; Goffman, 1963).

Third, and taking up these points about vitality and identifications, clubbing may not represent 'resistance' in the terms that Tariq Ali (1998) appears to mean it (see also Ali and Watkins, 1998). Yet this general lack of engagement with politics does not infer – and must not be confused nor conflated with – a general lack of engagement with 'the political' (Pile, 1997). I would strongly contest Ali's assertion that today's youth is 'dumbed down' (1998: 15). Clubbing may not change the world in the sense that previous 'youth movements' attempted (and there is scant evidence that any of these movements succeeded – thanks for nothing, parents!), but clubbing certainly temporarily and significantly alters the social worlds of those who enjoy it. Furthermore, the alternate, interactional orderings through which clubbing is constituted do impact upon the identities of the clubbers concerned, as well as being – by definition, through their very preference for clubbing – important identificatory experiences. The 'different form of revolution' that clubbing can represent for some is constituted through the immanent vitality of the crowd and the strong emphasis upon the 'here and now', through the entwined notions of movement and emotion, and especially upon the imaginative spaces opened out through excessive sensory stimulation[3].

In this sense, clubbing seems to be much less about rather rigid conceptions of 'resistance' and notions of authority that must be tirelessly battled and fought, and more about notions of gaining 'strength to go on' through the sharing of a crowd ethos, about fluxing between notions of (egocentric) identities and of (lococentric) identifications, and about finding a space in which – even if just for a fleeting instant – to forget oneself (Maffesoli, 1995; Pile, 1997). Pitched in the terms of the existing debates, 'The night out' suggests that if clubbers are 'resisting' anything (and this is by no means clear), then it may be other aspects of *their own lives* – aspects in which the experiencing of 'flow' is more problematic (Csikszentmihalyi, 1975a), the 'burden of identity' is less easily (and/or less beneficially) eluded (Bauman, 1995b), and the ecstasies of this unburdening – the 'going beyond' of the self – are less powerfully experienced, if at all (Laski, 1961; 1980).

Experiential consuming

[W]hen places of consumption themselves evoke the feeling that they are unreal, magical, impermanent, and inauthentic, then the very grounds for experiencing reality are shaken.

(Sack, 1992: 4)

Clubbing is an example of experiential consuming – a form of consuming in which nothing material is 'taken home', but which can nevertheless produce important memories, emotional experiences and imaginaries (remembered imaginations) that can be sources of identification and thus of vitality. In narrow terms, the product of clubbing (what clubbers 'purchase') can be understood to be the imaginative and emotional spaces that clubbing provides. These spaces are used by clubbers to think, to reflect upon oneself and one's life, but also to experience others (and 'otherness') in a form unlike that possible in most other social spaces of everyday life, and to inhabit a world that exists in the uncertain space between imagination and performance. The intensity of the clubbing crowd and the elective and affective identifications that can spin out of this intensity mark out the clubbing crowd as qualitatively different from many other forms of group togetherness.

First, the ways in which the clubbing experience is synchronously produced and consumed by the clubbers through a fusion of bodily practices, emotions and imaginations is as important as what it is that they are consuming. Thus, the clubbers' evocations and descriptions of techniques, skills, notions of competency, belonging, identification, and the sensations of freedom and of inhabiting an alternate, and in part imagined, world are partly what is being consumed in the clubbing experience (and I discuss this further on p. 185).

Second, it follows that theorisations of consuming which privilege moments of purchase alone are completely inadequate in providing effective understandings of clubbing. They provide only a meaningless 'snapshot' of purchase, little context for that purchase in terms of the processes involved prior and subsequent to that purchase, and reduce consuming choices to a prescribed and inflexible lifestyle affiliation. Clubbing is experienced through on-going practices rather than as a fixed 'state' which is simply inhabited. Clubbing is experienced through unfolding processes and negotiations rather than through simple notions of possessed properties. In being crowd-based, clubbing is experienced through continual dialogues and interactions, as opposed to soliloquies and solos. Clubbing is better evoked as a dynamic *process* of consuming rather than a fixed consumption relation, moment or state. Clubbing does not 'happen', but is experienced.

Third, while at first glance apparently chaotic and without 'rules', clubbing is actually heavily imbued with processes and practices of social and spatial ordering. These geographies of clubbing are never fixed, but rather always in a process of becoming; they are also always open to practices of subversion and re-negotiation. This ordering is present in clubbing in the form of the *sociality* that runs through and underpins social life more generally: the intricacies of style(s), the hierarchies of 'coolness', techniques of dancing, and the spacings of all the constitutive elements and moments of the night – the door and the dance floor, to give just two examples. Furthermore, it appears that, certainly for some women, this sociality of clubbing offers a more attractive, less intimidatory and potentially more liberatory experience than many broadly similar experiences such as 'pubbing'.

Fourth, and related to this point about sociality, conceptualisations of the processes and spacings constituting consuming crowds are beginning to become more prominent. However, understandings of how consumers actively produce – or at least assist in the production of – the experience which they simultaneously consume remain limited and sketchy. 'The night out' provides an example of how the different clubbing experiences of the clubbers are both selected and constituted through the tensions between notions of belongings and notions of differentiation. The rules, customs and rituals that together constitute the sociality of clubbing and provide the foundations for the practices of belongings and distinctions are also structured through spatial (and thus also temporal) orderings. This is most clearly exemplified in the case of dancing.

Fifth, these intricate social–spatial formations which constitute clubbing question apparently neat and tidy theorisations of 'new' forms of communality. While, as I noted in 'The beginnings', sociality has often been presented as the glue that holds temporary communities, such as those of the dance floor, together – in that what goes without saying makes the community (Amirou, 1989; Maffesoli, 1993a) – it is a mistake to conflate the communality of clubbing with simplistic assertions about sociality. It takes so much more than knowing the rules of a game to be able to play it effectively – practices are about more than 'knowing how'.

Clubbing is constituted through the skills, competencies and shared knowledges which only repeated participation can instil. Clubbing is not just about wearing specific clothing, but also the way that clothing is worn; not just about presence within a space at a certain time, but also the way in which this space is inhabited through one's use of the body; not just about dancing in an acceptable way, but also about dancing in an acceptable space at the right time and possibly with the right people.

Some have suggested that the sociality through which experiences such as clubbing are partially constituted is indicative of new or revived forms of communality premised upon notions of 'neo-tribalism', in which individuals are free to move between the many social groupings with which they wish to feel an affective identity (see for example, Maffesoli, 1995). Yet, this glosses over the usually demanding practical negotiations and learning processes that affiliation with these social groupings infers (Csikszentmihalyi, 1975a; Goffman, 1967). Belongings do not magically present themselves, but must be worked at, even for those for whom it would appear belongings arise 'naturally', such as the apparently 'cool'. In a similar vein, others have proposed that 'we now exercise relative choice in selecting the kinds of communities with which we would like to be involved' (O'Connor, 1997: 153). 'The night out' has clearly demonstrated that these choices about which nights are preferred or are possible are always relative. Individual clubbers will find certain clubbing crowds more amenable than others, with some clubbing crowds (literally) trading on their exclusivity, while others sell themselves as all-inclusive and without restrictions. The neo-tribes thesis (Maffesoli, 1995) is useful in beginning to understand the less rigid structurings to which moves beyond narrowly prescribed identities can give rise, but it must be qualified by noting how notions of 'freedom' are always contextual. Belongings and identifications are thus about much more than questions of cultural capital and social structure (Bourdieu, 1984; Warde, 1994a; 1994b). Belongings and identifications also involve and, crucially, *demand* competencies, techniques and an awareness of the timings and spacings – the contextualities – of sociality.

Imaginations and fluctuations

Self-expression is less a matter of mood, energy, practice, and pep pills than it is of feeling that the people around you are *with* you.

(Albert Goldman, 1978)[4]

The communality that can be experienced through clubbing arises through imagined notions of belonging and especially through associated fluctuations between a privileging of self and a privileging of the crowd. The spacings of clubbing, and

185

of dancing above all, are critical to these emerging notions of communality in a number of ways.

First, clubbers both perform but also actively imaginatively *construct* the clubbing experience in terms of their understandings of themselves, of their changing relationships with others within the clubbing crowd, and of the varying socio-spatial demands of the night.

The practical techniques of bodily control which a clubber deploys in clubbing are related to that clubber's understandings of himself or herself and the *meanings* that s/he is investing within those practices at that moment. What a clubber does with his or her body is inextricably tied up with why s/he imagines s/he is doing it – body and mind are inseparable. The oceanic and ecstatic sensations, which clubbers experience to varying degrees and at different times in varying ways as the climax to the night out, providing the clubber with a sensation of extraordinary euphoria and joy, are particularly good examples of this relationship between body and imagination (or emotion), between movement and meaning.

During my discussion of playful vitality in the fourth section of 'The night out', I noted how falling back on notions of *communitas* (Turner, 1969) or unicity (Maffesoli, 1995) in developing understandings of clubbing would be to *assume* a diversity of clubbers within that crowd. I argued that the temporary unburdening of identities that membership of the clubbing crowd can facilitate does not necessarily infer that those identities were dramatically different from each other to begin with. I also argued that the notion of *communitas* suggests a process of conformity and a social similarity that is much too neat to be applied successfully to clubbing crowds. That said, and as Dyer (1993) infers so succinctly, to some extent *it does not matter* whether the clubbing crowd is actually diverse in terms of identities or not. Rather, in one sense, what matters is that many clubbers *understand* the clubbing crowd, of which they are a part, to be diversely constituted, and they crucially find this understanding a rewarding and enriching experience. The imaginative construction of the clubbing experience, as being premised upon this inclusion of different identities in the establishing of a sense of communality, is thus equally significant to these same notions of communality as are the practices and spacings of the body. The practical and imaginative construction of clubbing is in this sense one and the same.

Second, the complex negotiations involved in clubbers' continually evolving understandings of style and coolness suggest that, as well as being about notions of *belonging*, membership of – or, put another way, identification with – a clubbing crowd can also be about notions of *differentiation*. Vitally, these differentiations are present both between but also *within* clubbing crowds. This further suggests that it is simplistic to assert that identities somehow magically dissolve within the identifications that the clubbing crowds can certainly provide, as the *communitas* thesis infers. Each clubber within a clubbing crowd will fluctuate between experiencing that crowd in terms of an identification and in terms of being a discrete person within it. At any one moment, aspects of an individual's identity or aspects of an individual's identification with the crowd will take precedence. One moment a clubber will literally be self-conscious while the next s/he may feel as if they have ceded control of their body and even of their mind to the clubbing crowd, the music and the atmosphere. This fluctuation between self and crowd is a defining feature of the clubbing experience.

The seemingly unreal, yet also extremely vivid experiences of clubbing can allow clubbers to go beyond themselves. Yet, and seemingly paradoxically, through this going-beyond they may find something more of – as well as something of extraordinary value within – the very self outside of which they may temporarily slip.

NIGHTS OUT

If disco is emblematic of where it's all at today, then the stunning profusion of lights, sounds, rhythms, motions, drugs, spectacles, and illusions that comprises the disco ambience must be interpreted as our contemporary formula for pleasure and high times. The essence of this formula is the concentration of extremes. Everything is taken as far as it can be taken; then it is combined with every other extreme to produce the final rape of the human sensorium. Why?

(Albert Goldman, 1978)[5]

This book represents only the beginnings of an understanding of the countless meanings, interpretations, practices, rewards and experiences of clubbing. The night out and the ecstatic geographies that I have traced through it are but tentative motionings towards a fuller conceptualisation of what and how clubbing can mean something significant to those who enjoy it. Through drawing out and developing notions and conceptualisations of 'play', of 'vitality', and of the complex relationships between clubbers' identities and their clubbing identifications, I have attempted to evoke the clubbing experience. I have also attempted to engage and progress certain existing literatures and understandings of the constitutive elements of clubbing: music, dancing, performance, crowds, communality. *Clubbing: Dancing, Ecstasy and Vitality* has started to provide some answers to the question posed by Albert Goldman twenty years ago: Why?

Clubbing is both extraordinary and ordinary. Clubbing is at once a *special* and yet also for many currently a *fundamental* facet of their lives. Clubbing is a blending of physical and emotional *movement*: the vitality and unspeakable feeling of strength that may be experienced through the drop of the bass, the first beats of a favourite track, a fleeting instant of belonging within a ceaselessly moving crowd, a welcoming glance of warmth from a stranger, a letting go, a going beyond, a simultaneous affirmation of self and of belonging through motion and emotion – through motion *as* emotion.

That was how it seemed to Glen. His pill was kicking in, and the music, which he had had a resistance to, was getting into him from all sides, surging through his body in waves, defining his emotions. Before it had all seemed jerky and disjointed, it was pushing and pulling at him, irritating him. Now he was ready to go with it, his body bubbling and flowing in all ways to the roaring bass-lines and tearing dub plates. All the joy of love for everything good was in him, though he could see all the bad things in Britain; in fact this twentieth-century, urban blues music defined and illustrated them more sharply then ever. Yet he wasn't scared and he wasn't down about it: he could see what needed to be done to get away from them. It was the party: he felt that you had to party, you had to party harder than ever. It was the only way. It was your duty to show that you were still alive. Political sloganeering and posturing meant nothing; you had to celebrate the joy of life in the face of all those grey forces and dead spirits who controlled everything, who fucked with your head and your livelihood anyway, if you weren't one of them. You had to let them know that in spite of their best efforts to make you like them, to make you dead, you were still alive. Glen knew that this wasn't the complete answer, because it would all still be there when you stopped, but it was the best show in town right now. It was certainly the only one he wanted to be at.[6]

DIONNE [An extract from *The Moment*[7]]: Arms in the air, not waving and whooping like a silly kid, just moving and feeling it all through my body. We look out over a sea of faces, pulsating colours, beautiful music. I'm smiling like a smiley thing...this is it...I hold my breath...this is the moment.

APPENDIX
Biographical snapshots of the clubbers[1]

Antonio

A 26 year old teacher who lives and works in London, Antonio particularly enjoys 'jazz-funk, easy listening and soul clubs', although he admits that he does not venture out into a wide variety of places but instead 'tends to play safe'. He notes how he is now going clubbing less and less regularly, instead picking out special nights very much for the DJs who are playing that night; he 'probably like[s] the idea about going to a certain sort of club as much as being there'. He volunteered in response to my letter in *i-D* magazine and is distinctive amongst the clubbers whom I interviewed in that he claimed to have never taken ecstasy (MDMA).

Bruce

Bruce is 26, a trainee producer, and lives in London. Bruce made contact with me as a result of my letter in *i-D* magazine. He has been clubbing for seven years and this interest has impacted upon his work, where he has 'made one short film about clubbing and [is] trying to get another financed independently'. Bruce claims that 'clubbing and its associated trappings has really altered my life. Not only by changing the music I listen to but it affects the clothes I wear, the people I see, the books I read and the way I feel about myself'. Bruce started clubbing in earnest while he was a student at university, and prefers clubs playing 'house' music. He particularly notes how he enjoys the post-clubbing part of the night when 'you quite often end up going home with strangers almost...so they'll be an interesting thing going on there and you'll have someone you don't know who's very off their face in your flat'. Bruce values his clubbing experience, greatly and currently has 'little intention of giving up'.

Carmel

Carmel is 25, works in London, and contacted me through the uk-dance discussion list. She lives in north London. Carmel was introduced to clubbing by a friend of a friend and was immediately struck by the quite different atmosphere she experienced during her first night out: 'the one thing I remember from there is a girl pushing past me and saying "I'm really sorry" – y'know everyone was so nice'. Although Carmel enjoyed mostly free, usually illegal, parties at the very start of the 1990s, she now also goes to 'house' clubs. Carmel is currently completely absorbed by clubbing and her free party memories, but she can envisage a time when she might not be 'so into the scene'. On a number of different occasions Carmel spoke about wanting to start a family and wanting to be friends with her parents again, about 'growing up and moving on'.

Carol

Carol is a 21 year old working and living in London. She contacted me through the uk-dance discussion list and enjoys a variety of clubs, but especially those that play mostly 'house' music – 'it's happy music, it gives you a bit of a buzz, it makes me feel happy'. Carol contrasts her earliest experiences with dance music at raves when she was 16 – 'we'd meet in a field' – with the 'proper clubs' that she has now been going to for about two years. This distinction between raving and clubbing is crucial for Carol – 'if I go out to a club then I don't say that I'm going raving, because I associate raving with people dressed in white gloves and their whistles [whereas with] clubbing you just get people who go out to enjoy themselves rather than people who just go out to go raving'. The music is central for Carol – she cannot understand, for example, why 'people go on about DJs' and describes a moment in one night where 'a DJ came on and we thought he might be quite good – everyone was like "Yeahhhh!", screaming and whistling and everything, but I thought it was crap'.

Dionne

Dionne contacted me after seeing my request in *The Face* magazine. She is 28 and works in central London. Dionne lives in London. She did not start going out properly until she went to university: 'my mum and dad wouldn't let me…I usually had to ask permission and they said "no"'. Before her interest in clubbing developed, Dionne was into live music – she has been to the Glastonbury festival every year since 1986 – and she has retained and developed this interest in parallel with her clubbing experiences. Her first experience of what she would call 'clubbing' came in 1988, although she did not come across drug use in the scene until the 1993 Glastonbury festival, where she took ecstasy (MDMA) for the first time: 'that's late compared to other people...the Summers of Love (1989 and 1990) just passed me by completely'. Although she enjoys clubbing very much, especially 'underground techno nights', Dionne also enjoys football, live events of any kind and cinema. She was the only clubber interviewee to discuss her other interests in some detail, and notably without any prompting: 'I think to have anything as the be all and end all of your life, whether it's work, clubbing...I think that it's very sad and if you can't be a balanced person with loads of facets in your life then you're really really sad'. Dionne enclosed a personal, yet very evocative short story – The Moment – with her reply to my request for volunteers.

Dwayne

Dwayne is a 28 year old working and living in central London. He contacted me through the uk-dance discussion list. Dwayne started clubbing at a relatively young age after promptings from friends and on hearing rumours of what clubs were supposed to be like: 'I was probably about 15 or 16 and my friends started going...I had always thought of clubs as being incredibly exciting, and places where I wanted to be...they were dark, they were loud, attractive girls. I found it immediately extremely attractive, ummm, and it was doubly exciting for me because at that age my parents wouldn't allow me to go out late so I had to climb out of the windows – my parents were very religious'. Dwayne now goes to mainly 'jungle and techno' clubs in central London. The 'coolness' and the music that constitute clubbing are important motivational features for Dwayne – he admits to being very picky about which nights he likes: 'I'd say, checking all of the clubbing experiences in my life, 90 per cent of them were not really good'. Dwayne adds that, partly because of his age, he particularly dislikes nights where there are lots of younger people.

John

John works for a City firm and volunteered to help through the uk-dance mailing list. John is 24, lives in Essex, and has been going out for eighteen months since a friend at work introduced him to London clubbing. He enjoys mostly 'house' and 'happy house' club nights, but is 'open to anything', and particularly likes 'meeting like-minded people – people who are pretty much the same as me'. John's first 'proper clubbing' experience – in central London as opposed to local clubs – coincided with his first experience of the drug ecstasy (MDMA): 'I became very aware of myself, aware that I was in different surroundings, different situation, different people, didn't quite know what was going on'. After an initial explosion in the number of times he went out during which he went clubbing and took drugs every weekend, John is now trying to 'have a month off every now and then'.

Kim

Although she lives in Essex and goes clubbing in central London, Kim is currently at university in northern England, where she also goes clubbing. Aged 21, Kim contacted me through the uk-dance discussion list. Like many other women, Kim draws distinctions between pubs and suburban clubs on the one hand and 'proper clubbing' on the other. Kim enjoys mainly 'house' music clubs and takes ecstasy (MDMA) on about 'half the times' that she goes out, although she had been clubbing for about six months before she first tried ecstasy (MDMA). At the moment Kim cannot see an end to clubbing, although she suggests that it will probably fade away: 'not completely, but it just won't be so regular…like every couple of months and then maybe like two, three times a year, something like that'.

Lucy

Lucy is close friends with Carol, with whom she goes clubbing regularly. She is 21 and works in administration in central London, where she also lives, and she contacted me through the uk-dance discussion list on the Internet. Lucy draws out what for her are important distinctions between clubs and pubs, and especially picks out the different sense of visibility and personal comfort she feels in each as significant in her enjoyment of clubbing: 'you get more locals down the pub, whereas at the club it's everyone isn't it…you walk in a pub and you're looked up and down [whereas] in a club it doesn't matter if you haven't been there before – you feel at home'. Lucy enjoys 'happy house' clubs, tends to go out 'in spurts' – 'sometimes I'll go out every weekend for two months and then I won't go out for like three months' – and, while she often goes clubbing with Carol, Lucy also goes clubbing with different groups of friends, sometimes just her boyfriend. In a similar way to Carol, Lucy is more impressed with the music itself than the DJs who play it. Lucy currently sees no end to her clubbing.

Luke

Luke is 21, works in publishing, and contacted me in response to my request in *The Face* magazine. As he lives in Holland, we had originally planned to meet in London and talk about his experiences in London but, while this never actually happened, we had developed an interesting discussion over the Internet and we decided that Luke might be able to recount an interesting set of experiences anyway. I 'interviewed' Luke twice over the Internet, although we exchanged mails much more frequently, and we were still in touch with each other in mid-1998. Luke started clubbing when he was 16, and once he had discovered 'mellow house' music this interest really took off: 'it was really one of the best summers I have had 'til now. Everything was new, I met lots of people, I really liked to go out'. Luke enjoys meeting people at clubs and will often go clubbing alone with the intention of making new contacts at the club. For Luke, the music 'gives me energy, quite literally', and combined with ecstasy (MDMA, or 'XTC' as Luke often referred to it in his e-mails), this energy 'makes [him] feel good, so [he] has good thoughts'. He adds: clubbing represents a 'very good way for me to forget my daily problems and other things I'm struggling with and a good night can give me the energy to continue to do things'. While he continues to enjoy clubbing as much as he presently does, Luke can see no reason to stop.

Maria

A 24 year old living and working in central London, Maria contacted me after seeing my letter in *The Face* magazine. Maria continually drew comparisons between Southend, where she was born and brought up, and London. She started clubbing much more often when she moved to London to go to university, although she found it very difficult to afford 'London clubbing' as a student. As part of her university course Maria lived in Amsterdam, where her interest in clubbing grew. Maria now goes most often to clubs playing 'funky sort of uplifting techno music…some hard house, some sorts of acid jazz [and] punk stuff', having tried other musical and clubbing genres and not liked them as much. Maria often takes ecstasy (MDMA) and the amphetamine speed when she goes out and this constitutes a significant aspect of her clubbing experiences: 'I'd say that 90 per cent of the time I will have an enjoyable evening, whatever the music – I think that has to be something to do with the drug'.

Mark

Mark is 26. Although he lives in Slough, Mark works for a central London-based consultancy – a fact which he finds continually amusing in an 'if only they knew!' fashion. Mark responded to my request for assistance made on the Internet discussion list uk-dance, and is 'crazy' about the Internet: he has his own web-site on which he advertises club nights he likes. Mark uses the uk-dance discussion list in making new contacts who 'share an interest in trance music' for, as he explains, 'there's not a lot of trance going on in Slough, as you can imagine'. While Mark's interest in clubbing was triggered by his arrival in (or near) London, post-university, his fascination with clubbing – and especially his love of 'Goa' trance nights – was prompted by travelling experiences in Japan and Thailand. Mark has also travelled to India on numerous occasions, mainly for the Goa beach raves around Christmas.

Sarah

A 26 year old nutritionist, Sarah lives and works in north-west London. She responded to my letter in *The Face* magazine requesting volunteers. Sarah started clubbing when she was 16, through a mixture of 'peer pressure' and 'curiosity', and now goes clubbing about once a fortnight. Although Sarah goes to a variety of different clubs she tends to favour one particular club over all others: 'the first time...I just walked through the door and it was like...excuse my language, but it was just like: "headfuck" – this is the place for me...it was totally different to that other place that I had been to and I knew that this was the place that I was looking for'. Sarah enjoys meeting friends at the club, which she describes as a 'glorified pub' for this reason, but is currently finding herself going to the club, and to other clubs, much less – her friends she met there have become more important than the club itself.

Seb

Seb was quite unlike the other clubber interviewees in the intensity of his almost fanaticism for clubbing. An extract from his initial letter, in response to my letter (in *The Face* or *i-D*, it is unspecified), provides some personal details and reads as follows:

Dear Ben

Do I enjoy clubbing in London?...Sit back, relax, and let me tell you how much!

My sister wrote to me having read your letter. Quite simply she and I are of the same opinion – if you're looking for help/information to complete your thesis, you need look no further. I'm 26 and have been clubbing in one form or another for the past five years. In that time I've travelled from Aberdeen to London and taken in Glasgow, Leeds, Sheffield, Manchester, Liverpool and other places in search of a good time. As you can see I currently live in Liverpool, though I'm trying to find work in or around London, but this doesn't stop me travelling to London two/three times a month for some exceptional parties. I've got a lot to say on the subject of clubbing – probably more than your average clubber. Indeed, I intend to write a book on the subject at some point in the future and have been collecting 'stories' to include in that for the past couple of years. If you want to know why I go clubbing, the answer's simple – because I can! (though I'd be happy to elaborate on this answer if you want me to!).

...

I'll leave things at that for you now though I could go on for hours...if you think I could help at all, I'd be delighted to hear from you.

Yours in anticipation

Seb

Simon

Simon is 27. He lives and works in central London. He became interested in clubbing while at university and has been clubbing regularly since he was 20. He particularly likes 'jungle and techno' nights, although he also frequently goes to clubs playing 'Asian dance music and jungle'. Simon's use of drugs such as ecstasy (MDMA) varies between nights – while he would 'hardly ever take drugs at a drum'n'bass night, mainly because I wouldn't see it as an essential part of it', he would more usually take some form of drug – usually ecstasy or the amphetamine speed – at 'techno' nights. He spent large sections of the interviews talking in detail about all aspects of the music involved, and was one of two clubber interviewees who had DJ-ing experience that they told me about.

Sun

Sun is a 25 year old journalist who works in central London. He came to be into clubbing over a period of time rather than as the result of a single night out: 'it's a bit like losing your virginity in that people say it happens overnight but actually it fades in over a period of time'. Sun enjoys most clubbing genres with the exception of trance-based clubs, which he hated, but describes himself as a 'jungle fanatic'. Sun's parents moved to London from India when he was much younger – a feature of his early life that undoubtedly comes through in the three conversations we had together. Sun responded to my request for assistance made on the uk-dance discussion list.

Valerie

Valerie is 22. She used to be an actor and a lecturer in Drama. She contacted me through the uk-dance discussion list after my request for volunteers was drawn to her attention by Kim, her close friend. Both Valerie's parents are Greek and she has travelled to Greece for a month every year throughout her life. She works in central London. Valerie's favourite music is 'happy hard-core, hard-core, ambient trance, Goa trance, garage and many others depending on my mood'. Valerie enjoys one particular club above all others – it provides a 'guaranteed good night out', which she feels is important.

NOTES

1 In this book all clubbers' names, club names and club nights are pseudonyms.

2 This night out is told retrospectively from notes made during the night itself and immediately afterwards. These notes were fleshed out in detail during the following morning. I briefly discuss how I conducted the research on which much of the book is based at the end of this set of introductions.

PART ONE: THE BEGINNINGS

1 For further discussion of clubbing as a cultural industry, see also: Mintel, 1996; Thornton, 1995; and O'Connor and Wynne, 1994.

2 Mintel (1996) interviewed a 'nationally representative sample' of 1,087 adults during late April and early May 1996. Mintel reports are expensive enough (£450 for the *Nightclubs and Discotheques* survey) to be far beyond the means of individual researchers. However, many business libraries, such as City Business Library in central London, hold full sets of these and other similar reports.

3 Release is an independent drugs advice and research organisation that is mainly concerned with the health, welfare and civil rights of those who come into contact with drugs. Their Dance Outreach Network provides information and support at clubs, festivals and other dance-related events. Release conducted face-to-face interviews with clubbers 'at eighteen venues throughout London and the South-east between March and November 1996' (1997: 2) using a blend of structured questionnaires and further interviews. Release's respondents were selected on 'an anonymous randomised basis with confidentiality guaranteed' (ibid.).

4 These new superbars offer unprecedented choice within a single venue. One of the central London superbars offers seven different bars, three dance floors, three restaurants, a take-away, a chill-out area and a games area – all accessible without encountering either the English winter or yet another queue.

5 I discuss the use of drugs as part of the clubbing experience in the third section of 'The night out'. Thornton (1995) offers a brief history of the development and nature of the Ritzy discotheque.

6 Where possible, I use the term 'young people' throughout the book in preference to the term 'youth', with the exception of where it is necessary (i.e. in discussions of the

literatures of the 'youth subcultures'). For me, the term 'youth' resonates with a slightly unsettling, de-humanising, patronising and ascribed quality.

7 Perhaps the epitome of this transition to an image of trouble-as-fun, fun-as-trouble (Hebdige, 1983: 401) were the clashes between mods and rockers in Brighton and Margate in the early 1960s.

8 In addition, the early studies of young people and the 'resistance' of the CCCS provided the foundations of the fledgling discipline of Cultural Studies in Britain.

9 See, for example, the accounts presented by Goldman, 1978; Goldman 1993; Haden-Guest, 1997; Hughes, 1994; and Rietveld, 1998b.

10 For example, Gibson and Zagora (1997) and the Rave Research Collective, with which they are associated (a collective of academics and non-academics, clubbers and musicians), provide an understanding of 'rave culture' as it exists in Sydney. They are particularly interested in the social and cultural contexts from which the 'spaces of rave' are emerging, and the processes and practices through which diffusion and commodification have diversified and challenged the definitions of the Sydney rave scene (Gibson and Zagora, 1997). Details of the Rave Research Collective and the recently launched (spring 1998) Youth and Clubbing Discussion List can be obtained from Chris Gibson by e-mail (cgibson@mail.usdy.edu.au). For further recent contributions towards understandings of clubbing cultures, see also: Wright (1995; 1997; 1998) on the notion of dance cultures as a symbolic social movement and the roles played by gender, class and ethnicity within these dance cultures; Gilbert (1997) on the implications of the Criminal Justice Act upon the British dance/rave scenes; Jordan (1995) for a Deleuze and Guattari-inspired theoretically centred discussion of raving; Chapter 6 of Collin (1997) provides an excellent summary of the growth of the 'techno traveller' culture from its earliest beginnings in the free Stonehenge festivals of the 1950s and 1960s. Also see Gore (1997); O'Connor (1997) and Pini (1997a; 1997b) on the specific experience of women within raving and clubbing cultures; Merchant and MacDonald (1994) on the diversity and democracy of youth subcultures more generally, including dance cultures; Hemment (1997) on both philosophical and musical understandings of 'ekstasis' and on the role of the DJ in dance cultures; Urquía (1996) for a sideways swipe at the role of DJs at dance events; Saldanha (1998) provides an ethnographic approach to understandings of young urban elites in Bangalore (India) and the socio-cultural tensions brought to the fore by their clubbing practices; and, recently published work by Rietveld (1998a; 1998b) on house music cultures in the United States, Britain and Holland. Simon Reynolds' *Energy Flash* (1998) provides an authoritative and comprehensive account of the development and transformation of rave culture over the last ten to twelve years, while Sheryl Garratt's *Adventures in Wonderland* (1998) provides a superbly detailed account of the birth of rave and club culture in 1988 and 1989 especially.

11 The sub-title 'shut up and dance' has two sources: most pertinently, it is the title of an important chapter (1994b) in McRobbie's (1994a) text on youth cultures and cultural studies in which she discusses 'rave'; less obviously – and one wonders if this is where McRobbie gleaned the phrase – 'Shut Up And Dance' was also the name of an early 1990s rave group.

12 In the British context the most obvious youth-based cultural forms that may indeed have *mega*-political intentions as at least part of their *raison d'être* are the travelling cultures documented by Hetherington (1992), Halfacree (1996) and McKay (1996).

See also 'Uncivil Societies' in the special issue of *New Formations*, spring/summer 1997, vol. 31.

13 Where possible, the term 'consuming' is used in preference to 'consumption'. In a similar way to which I have been stressing the process rather than the state of being young, consuming evokes the *process* through which all consumption is *practised*, while consumption too often suggests merely an *act* of purchase or a *moment* of transaction. I deepen this point shortly.

14 See, for example, the neat summary of various approaches and understandings of consumption by Bocock (1993); Clarke's (1996) chapter questioning the current limits of retail geographies; Jackson's (1989) seminal contribution to 'new cultural' understandings of consumption and identity; Kowinski (1985) on shopping malls; Lee (1993) on the changing cultural politics of the 'consumer society'; Lunt and Livingstone (1992) for a more psychological approach to relationships between consumption and identities; McCracken (1988) for a detailed review of late 1980s understandings of studies in consumption; McRobbie (1994a) for her explicitly youth-orientated 'take' on the place of consumption within understandings of popular cultures; Miller (1989) for a more anthropological approach; Mort's (1989) and Nava's (1992) more Cultural Studies-style understandings; Sack's (1992) treatment of the relationships between notions of place and the ways in which we experience – and consume – the world around us; Shields (1992a) on the emergence of 'new identifications' through lifestyle and consumption cultures; Tomlinson (1990) on consumption and advertising; and Wrigley and Lowe (1996) on aspects of retailing and their links with consumption. Two special issues of *Environment and Planning A* (1995; vol. 27) on 'Changing geographies of consumption' splice the 'new cultural turn in human geography' with the emergence of a 'new retail geography' in attempting to historicise, contextualise and empirically ground geographies of consumption (Jackson, 1995: 1875).

15 These commentaries upon commentaries include: Edgell, Hetherington and Warde (1997); Friedman's (1994) edited collection on geographies of consumption and identity; Crewe and Lowe (1995) and Gregson and Crewe (1994; 1997a; 1997b) on alternative understandings of shopping and its role in constituting identities; Danny Miller's (1995) edited review of new studies of consumption across different social science sub-disciplines; and Sulkunen, *et al.* (1997) on new approaches to what they refer to as 'contemporary consumer society'.

16 On geographies of the shopping mall see, for example: Chaney (1990); Crawford (1992); Goss (1992; 1993); Hopkins (1990; 1991); Jackson and Johnson (1991); Kowinski (1985); Morris (1988); Shields (1989; 1991a; 1992a); Sorkin (1992); and Williamson (1992).

17 Contributions towards a social history/sociology/geography of the department store include Abelson (1989); Blomley (1996); and Dowling (1991).

18 A number of writers have looked at the fairs, centennials, exhibitions and carnivals held in both European and North American cities, for instance: de Cauter (1994); Greenhalgh (1988); Jackson (1988); Ley and Olds (1988); and Pred (1994). The imminent millennium celebrations will surely trigger a further rash of similar works (see, for example, Pinnegar (1999), forthcoming).

19 On consuming food more generally see: Acre and Marsden (1993); Bell and Valentine (1997); Cook (1995), Cook and Crang (1996); Mennell, *et al.* (1992); and Warde (1996).

20 Still curiously understudied, analyses of the consumption-advertising interface have expanded considerably in recent years (Nava, *et al.*, 1997). See, for example, Goldman (1992); Lury and Warde (1997); the recent collection edited by Nava, *et al.* (1997); Williams (1993 [1962]); and Williamson (1978).

21 An increasingly significant exception to this paucity is work by Nicky Gregson and Louise Crewe on car-boot sales, thrift and charity shops (Crewe and Gregson, 1998; Gregson and Crewe, 1994; 1997a; 1997b; see also Gregson, Crewe and Longstaff, 1997).

22 See Glennie and Thrift (1996) for further discussion. The current formulations they discuss include those of Hetherington (1996), Maffesoli (1988b; 1993a; 1995 [1988]; 1997) and Shields (1991b; 1992a; 1992b; 1992c; 1995).

23 Michel Maffesoli is Professor of Sociology at the Sorbonne, University of Paris V, and Director of the *Centre d'Etudes sur Actuel et le Quotidien* (trans. Centre for the Study of the Real and the Everyday). Major publications include: *The Shadow of Dionysus: A Contribution to the Sociology of the Orgy* (1993a), *The Time of the Tribes: The Decline of Individualism in Mass Society* (1995), *Ordinary Knowledge: An Introduction to Interpretative Sociology* (1996a) and *The Contemplation of the World: Figures of Community Style* (1996b). These have all been previously published in French (in 1985a; 1988a, 1985b and 1993b respectively). Maffesoli's work is increasingly finding its way into understandings of contemporary societies, for example, see: Bauman (1992; 1993; 1995a), Glennie and Thrift (1996), Hetherington (1994; 1996; 1997), Lash and Urry (1994); Lovatt (1996); Tester (1993); Touraine (1995); and Wright (1997). A critical evaluation of the work of Maffesoli can be found in Evans (1997), and a shorter engagement can be found in Crang and Malbon (1996).

24 I discuss the notion of 'identifications' in more depth in the first section of 'The night out' (p. 48).

25 This point spins out of much larger, more enduring and more detailed debates over inter-subjectivity, and specifically out of critiques of Cartesian assumptions that mind and body are in some way independent of each other. Crossley (1995) provides a detailed yet succinct overview of this debate, discussing the notion of subjectivity through a fusion of the concepts of 'body techniques' (Mauss, 1979) and 'intercorporeality' (Merleau-Ponty, 1962), with a radical re-reading of Goffman's *Relations in Public* (1971).

26 A full treatment of Butler's approach is beyond the bounds of this book, but is covered elsewhere. For further discussion, evaluation and critique, see: Aalten (1997), Cream (1995), Longhurst (1995; 1997), Mahtani (1998), Pinnegar (1995) and Rose (1995).

27 Conceived of in this way, Bourdieu's (1984) concept of the 'habitus' might be utilised in developing understandings of, for example, how notions of 'coolness' impact upon the practical, embodied techniques one employs in social situations in attempting to communicate or re-state that coolness. The practices of consuming are — to varying extents — pre-scripted, if only in some cases by prior experiences. I develop these points in sections one and two of 'The night out'.

28 A much fuller discussion of the methodological implications and potential problems of researching clubbing may be found within Malbon (1998).

29 The two Internet discussion lists of which I made particular use were 'uk-dance' and 'alt.rave'.

30 Brief biographical sketches of each of the clubber interviewees can be found in the Appendix.

PART TWO: THE NIGHT OUT

1 I prefer not to use the term '*flâneur*' in the *strict* sense that Walter Benjamin (1973) employs it in his narration of Baudelaire's notion of the 'man of the crowd'. Rather, I use the term more simply to evoke the mentality of those who suggest they can feel at home amongst strangers, and I, of course, include both women and men in this evocation.

2 A 'gabber' is clubber slang for a very hard-core clubber who has a penchant for exceptionally fast and hard techno music. The use of the term seemed to peak in 1996 and early 1997, and, at the time of writing (late 1998), is rarely heard in everyday clubber use.

3 This idea is also reminiscent of Georg Simmel's (1950) notion of the 'blasé outlook' that he claims characterised the social life of the contemporary metropolis (Simmel was writing during the early years of this century). This 'blasé outlook' was premised upon an attitude characterised by 'an indifference toward the distinctions between things' – an indifference that can be engendered through the quick-fire stimulatory experience of crowds. Citizens, suggested Simmel, became de-sensitised to the shock, surprise or stimulation of the experience of differences.

4 For further discussion and comparison of the different appeal of clubs versus pubs, see Thornton (1995: 20–22).

5 For accounts which, to varying extents, appear to overstate the 'liberatory potential' of clubbing for women see: Gore (1997), Gotfrit (1988), Griffiths (1988), McRobbie (1991; 1994b; 1994c), O'Connor (1997), and Pini (1997a; 1997b). It should be stressed that each of these accounts provides incisive and interesting discussions of dancing and clubbing more generally, and especially with reference to women's experiences. What I argue is an over-stress on the purely liberatory potential of clubbing should not detract from this.

6 These debates over clubbing (or raving) and notions of resistance are further opened out in the fourth section of 'The night out'.

7 As I go on to discuss in the fourth section of 'The night out' this experience of clubbing, representing for the clubber an alternate experience (or persona) of themselves, can itself be a source of individual vitality – a powerfully rewarding experience.

8 These clearly complex relationships between sex, sexualisation, gender and clubbing would appear to demand further, and especially empirically grounded, research. While potentially fascinating, the development of a cross-cultural and cross-spatial understanding of gender roles and interactional orderings is beyond the bounds of the current project.

9 Jungle music was, at the time of this interview (1995), a relatively fresh genre of dance music involving hyper-syncopated beats that can run anywhere up to 160 bpm (Wright, 1995) and, although initially dismissed as a passing fad, has subsequently influenced many other genres of dance music (Bat, 1997; James, 1997).

10 It is a moot point as to whether these great collective social identities *ever* gave us the code of identity in any case – this may be just what we told ourselves about ourselves (Hall, 1991).

11 The term 'ecstasy' derives from the Greek term, *ekstasis*: a displacement, a trance, a going beyond. Interestingly, the drug ecstasy – also known as MDMA, its chemical abbreviation – was originally labelled 'empathy'. The brand name was only changed to 'ecstasy' because it was thought that it would sell better (Collin, 1997).

12 For a more detailed discussion of some of the ways in which occupational status can be linked to tastes in music, see Peterson and Simkus (1992).

13 These extensive and on-going debates over the nature and role of lifestyles in contemporary consumer society require a fuller treatment than is possible here. For further discussion, see Colin Campbell's (1995) sizzling attack on the sociological relevance and construction of lifestyles; David Chaney's (1996) comprehensive review of many different theorisations; and Kevin Hetherington's (1994) argument that the social centrality of certain sites in lifestyles suggests that they might be better understood in terms of 'communion' or (after Schmalenbach, 1922) as *Bünde*.

14 Thornton (1995) deals with these relationships between the 'hip' and the 'mainstream' in detail.

15 Guides include the bi-annually produced *UK Club Guide* and the monthly *Guest List*.

16 This is an extract from the clubbing journal from my first night out with John (08.03.96).

17 A clubbing maxim that was widely used in the early 1990s; it is now often tagged onto the signature of an e-mail.

18 I return to these points in the final section of the night.

19 A large club in south London, often attracting in excess of 2,000 clubbers on Friday and Saturday evenings.

20 'Blim burns' result from burning fragments of cannabis falling out of 'joints' onto a reclining smoker's chest – in this extract Maria possibly mentions them to stress the apparently dishevelled appearance of her friend in contrast to the 'office workers'.

21 Of course, it should not need stressing that in practice these facets are indistinguishable.

22 Noteworthy exceptions to this general silence include: Blacking's (1973; 1987) now classic anthropological studies; Canon's (1993) short paper on so-called 'world' music; Chamber's (1985) glide through the pop music and pop cultures of a youthful postwar Britain; Cohen's (1991a; 1991b; 1995) work on relationships between the city, local music cultures and regeneration; Finnegan's (1989) similar work on the myriad of musical cultures in a single town; the connections between music and notions of utopia drawn out by Flinn (1992); Frith's (1978; 1987; 1988; 1992; 1995; 1996) hugely influential and still-expanding *oeuvre*; Gibson and Zagora's (1997) and Halfacree and Kitchin's (1996) analysis of local music scenes through a geographical lens; Hollows and Milestone (1998), who very specifically set out to spatialise and contextualise the links between musical cultures, identities and places; work on music and regeneration in northern England by Hudson (1995); Lily Kong's (1995a; 1995b) cultural politics of music in Singapore; notions of youth and identity through music in Kruse (1993); Leppert and McClary's (1987) edited collection which covers a broad spectrum of themes relating music to society; Leyshon, Matless and Revill's (1998) collection of essays examining the relationships between music and notions of place and identity; Norris' (1989) edited collection of work on music and culture; the innovative and diverse edited collection of cultural studies of music and youth found in Ross and Rose (1994); Rouget's (1985) seminal work on music and the trance state; Said's (1991) thesis on connections between music, politics and culture; work on the

relationships between music and society in Shepherd (1991); Smith's (1994; 1997) arguments in favour of a geographical and cultural approach to understanding music; Storr's (1992) philosophical and psychological study; Toop's analyses (1991; 1995) of rap and ambient music; and Whiteley's (1997) edited collection exploring the connections between sexuality, gender and forms of popular music.

23 On the neglect of the aural within the social sciences, see also Fornäs (1997), Jackson, M. (1989), Polhemus (1993) and Stoller (1989).

24 The potential of the Walkman in permitting, 'the possibility, however fragile and however transitory, of imposing your soundscape on the surrounding aural environment and thereby domesticating the external world' (Chambers, 1994: 51) has also been noted elsewhere by Chambers (1990) and Chow (1993; 1994), who suggests that '(i)n the age of the Walkman...the emotions have become portable' (Chow, 1993: 397). See also the recent Open University text entirely devoted to cultural studies of the Walkman and in which the contributions on p. 77 are discussed and extracts reprinted (du Gay et al., 1997).

25 For critiques of Adorno's approach, see Gendron (1986), Malm and Wallis (1993) and Said (1991).

26 For further discussion of the relationships between musical cultures and identities, see Smith (1994; 1997). Smith places the musical at the centre of the geographical imagination, and illustrates her argument with examples of the role of music in empire building and in the creation of local identities and identifications in the midst of local global tensions. See also Cohen (1991a; 1991b) on rock cultures in Liverpool; Grossberg (1984) on rock and notions of resistance; Kruse (1993) on subcultural identities and music; Malcolmson (1995) on links between 'disco dancing', local identities and ethnic tensions in Bulgaria; and Regev (1997) on world music, globalisation and fields of cultural production.

27 Later in 'The night out' (Sections Three and Four) I explore some of these notions of 'other places', 'fantasy spaces' and 'utopia'.

28 Again, I discuss these issues around the addressing of social differences later in 'The night out' and in 'Reflections', which forms Part Three.

29 These connotations of dancing with primitivism have been a recurring feature in academic accounts more widely, where bodily adornment and differentiation have been viewed as demonstrating 'modernity' and 'sophistication', while bodily expression through mimicry and group affiliation has been constructed as 'primitive' and of the 'lower strata' in society (Simmel, 1997: 190).

30 Hanna (1987) outlines the development of the anthropology of dance, stressing both its early faltering and somewhat half-hearted nature, as well as its current vibrancy. Dance and dancing are currently undergoing a revival in terms of their study within the social sciences. The collection edited by Desmond (1997) provides a broad cross-section of new and recent work on dance within cultural studies; Foster's (1996) edited collection provides a broad and extensive summary of current work in dance studies and choreography; and Thomas (1995; 1997) reviews recent work and directions within sociology.

31 A 'trainspotter', at least in clubbing cultures, is someone who makes a point of and gains pleasure through identifying the multifarious details of tracks being played by the DJs (artist, remix, label, date, even bpm) – a form of expertise which, while many joke about it, is much respected within certain forms and genres of clubbing.

32 'E' is shorthand for the drug, ecstasy (MDMA). The practices and spacings of drug use are discussed in the following section of 'The night out'.

33 Notions of backstage and front stage are always relative – while the loos may be backstage for the clubbers, for the loo attendant the loos are frontstage, and it is only by leaving and wandering through the club *beyond* the loos that the loo attendant can experience any notion of being backstage, can go 'off duty'. Furthermore, even within the loos there will be degrees of backstage, with the cubicles even further backstage than the sinks/urinals. Finally, it should not need pointing out that even within the backstage regions – in this case, of the loos – many clubbers will 'keep up appearances' and maintain their style-centred performances of, say, the dance floor.

34 '[T]he ways in which from society to society men [sic] know how to use their bodies' (Mauss, 1979: 97).

35 Referring to the work of Hall (1955) on cross-cultural spacings, of Hediger (1955) on animal spacings and of Sommer (1959) on personal spacings, Goffman notes how, in the absence of physical barriers to the perimeters of their engagement (such as walls), people engaged within a social situation frequently attempt to create a barrier through the spacing and orientation of their bodies. Dancing clubbers will often orientate their bodies away from the open, and perhaps still filling dance floor, and towards either each other and/or the location of the DJ's mixing desk – often on a raised platform at the 'front' of the club. This point is evocative of Valerie's earlier remark about dancing in a circle.

36 This notion of an 'away', somewhat paradoxically experienced as a temporary annihilation of everything *beyond* the here and now, begins to problematise certain overly neat dichotomies between presencings and absencings, here and there, proximity and distance. I discuss this further in the next section of 'The night out'.

37 See also Simon Frith's (1996) discussion of the relationships between music and time.

38 As suggested earlier, accessing and representing these views of dancing, as 'the researcher', can be problematic. When the dancer reflects on the experience in order to recount it the dynamism of the experience is gone (Csikszentmihalyi, 1975b).

39 In one of the first uses of the term in evoking this sensation, Sigmund Freud (1961) compared the oceanic experience to the heights of being in love – a state where the boundary between ego and object threatens to melt away (Storr, 1992). For Freud, this melting boundary between self and other was interpreted as a temporary regression to infancy, where the infant at the mother's breast had not yet differentiated itself from the external world. Thus, for Freud, the oceanic represented a regression to a total merger with the world (Storr, 1992). Inglis (1989) attributes the first use of the term 'oceanic' to Jung (1957) in describing the experience of transcendence (see also Jung, 1973).

40 As I have already mentioned and as I will suggest in more detail, I am using the term 'oceanic' to refer to *all* so-called altered states experienced while clubbing (both drug induced and non-drug induced) in preference to the term 'ecstatic' because of the current connotation of the word 'ecstasy' with the drug MDMA. I am especially concerned to avoid inferring that all clubbing experiences in which sensations of altered perception are experienced involve the use of drugs such as ecstasy (MDMA). As I discuss later, this is not the case.

41 This striking testimony is that of a 16 year old girl who participated in the 'Questionnaire Survey' upon which Marghanita Laski partially bases her first text. The interview took place in the late 1950s; over forty years ago. The girl is

responding directly to the question: 'How would you describe the sensation of ecstasy?'.

42 Wyndham is describing his impressions of a 1920s London dancehall. The absence of violins and 'the attenuated skirts', and the presence of drugs, such as ecstasy (MDMA), and much louder music are the only significant differences over seventy years later.

43 All of Laski's sixty respondents underwent their 'ecstatic experiences' whilst alone. This is the complete opposite of instances of the oceanic in clubbing, where the crowd is a foundation of, and even a necessary pre-condition for, the experiencing of oceanic sensations.

44 A point I further discuss in both the final section of 'The night out' and in 'Reflections' that form the final part of the book.

45 Inglis (1989) notes how this ineffable quality of ecstatic and oceanic experiences has led cynics to ask why, if the experience was really as significant to the individual experiencing it as they said it was, could they not describe the experience coherently to others? Inglis suggests that these critics should pause and ask themselves whether they could 'describe sight "coherently" to a blind man, or the feelings generated by the sight and smell of a flower even to somebody familiar with them' (ibid: 262).

46 As in Valerie's description, in the previous section on dancing, of being 'forced to dance...I wasn't thinking about it, it was just happening to me'.

47 I refer to the *drug* ecstasy as 'ecstasy (MDMA)' throughout. This admittedly very clumsy and perhaps even annoying description is used in order to avert any danger of misunderstanding between the *drug* 'ecstasy (MDMA)' and the *state of being in* ecstasy. This is a critical distinction.

48 Where a 'regular user' was deemed to be someone who had taken ecstasy (MDMA) in the previous month and an 'older person' was deemed to be someone aged 30-plus.

49 On media and moral panics about drug use see: Bellos (1997); Campbell (1997); Daily Star (1997); Desenberg (1997); Perri 6 (1997); Saunders (1996); and Travis (1997).

50 MDMA is a member of the MDA family and only one of many derivatives (including MDMA, MDEA, MMDA, MBDB and MEDA), all of which are chemically synthesised from the oils of natural products, such as sassafras (ISDD, 1996). MDEA and MDA are two common drugs which are similar to MDMA and all three have similar effects, although the choice of 'connoisseurs' is reputedly MDMA because of its enhanced empathic effects (Saunders, 1997; Wright, 1999).

51 Some so-called 'ecstasy' pills contain only 'speedy' drugs such as caffeine, amphetamine and ephedrine, while others contain completely different drugs altogether. Worst of all are those containing dangerous and sometimes mildly poisonous fillers (Saunders, 1997). The Dutch drug research institute, Jellinek, tested 545 samples that were sold as 'ecstasy' in 1994. On average 83 per cent contained ecstasy (MDMA)-type drugs and 5 per cent contained paracetamol or unknown drugs (Saunders, 1997).

52 For more general histories on the use of drugs in attaining altered and ecstatic states, see also Berridge and Edwards (1981); Goodman *et al.* (1995); and Shivelbusch (1992).

53 'Adam' was an early street pseudonym for ecstasy (MDMA): 'the name "Adam" for MDMA is related to the innocent man as in the Garden of Eden – "being returned to

the natural state of innocence before guilt, shame and unworthiness arose"'
(Saunders, 1995: 15).

54 See, for example, the front page of *The Daily Star*, 29 October, 1997.

55 An extract from the short story, *The Moment*, written during 1996, which Dionne enclosed with her original letter replying to my request for clubbing volunteers in *The Face*. She adds in the letter that the story is autobiographical.

56 An account from a 'fisherman on the pier' when asked why he had gone to Margate and what he had got out of it (from a BBC TV broadcast in the *Man Alive* series on 3 October 1974; cited in Laski, 1980: 156). From this extract it is apparent that this man has a relationship with fishing that is not dissimilar to the relationship of clubbers with clubbing. This, to me fascinating, point is tackled in detail in the next section of 'The night out' where I discuss notions of play and escape in relation to the 'oceanic'.

57 This extract is from a 'clubbing journal' which, for a period, Seb kept in intricate detail. As perhaps the key member of a loose gathering of clubbers calling themselves the 'Party Posse', all of whom used to go to the same clubs (and who often hired coaches to travel to clubs in other cities), Seb took on responsibility for writing, copying and distributing a three or four page newsletter/report of each of their big Friday and Saturday nights to up to fifty clubbers who were there. This entry is extracted from a much larger narration of a night out he spent with friends.

58 See, for example, the edited collection by Keith and Pile (1993) in which relations between power, identity and resistance are addressed through the language of their spatialities. For discussions of evolving feminist perspectives on the relationships between spaces, places, identities and notions of power see: Bell *et al.* (1994), Dwyer (1998), Madge *et al.* (1997) and Valentine (1993).

59 Where, stated simply, power is generated through strategies, and resistance to that power is generated through tactics (Pile, 1997): 'while strategies define a territory marked by an inside and outside, resistances cross these spaces with "other interests and desires that are neither determined nor captured by the systems in which they develop"' (Pile, 1997: 15; citing de Certeau, 1984: xviii). While 'political, economic, and scientific rationality [have] been constructed on [the] strategic model... many everyday practices (talking, reading, moving about, shopping, cooking etc.) are tactical in character' (de Certeau, 1984: xix).

60 Tim Cresswell makes this notion of 'intentionality' and 'being noticed' central in his distinction between 'resistance' and 'transgression': 'transgression is judged by those who react to it, while resistance rests on the intentions of the actors', although he goes on to add how these are ideal types, and, for example, 'some acts judged as constituting transgression are intended by the actors and thus also constitute resistance' (1996: 23).

61 A Latin term meaning to 'enjoy the pleasures of the day, without concern for the future...literally: seize the day!' (Collins Concise Dictionary, 1995).

62 The term *puissance* is derived from Old French, and etymologically means 'mighty' or 'having power'. The translator of *The Time of the Tribes* (1995) translates *puissance* as 'the inherent energy and vital force of the people' (ibid.: 1).

63 This extract is from another one of Seb's clubbing journals and is reproduced as in the original – the emphases are his.

64 This important point is discussed further, in the context of broader notions of communality and 'resistance', in the next and final part of the book, 'Reflections'.

65 From the interview with Bruce (see pp. 161–2).

PART THREE: REFLECTIONS

1 I use the term 'afterglow' from Laski (1961).
2 I will refer to these complex practices and spacings of this post-clubbing period in only brief detail for, as is the case with the pre-clubbing 'rituals' of scoring drugs, shopping for clothes or meeting in bars, a full engagement is beyond the confines of the current project. A short film entitled *Coming Down* (1997) and the accompanying soundtrack – both of which are directed and produced by Matt Winn – beautifully capture the mixture of reflection, excitement, euphoria, exhaustion, denial and the subtly different forms of group interaction and individuation which can characterise this period of the night. Also, see further points (p. 178) under 'Other lives, other identifications – clubbing in the city'.
3 For counter-arguments to this notion of inwardly directed politicisation, see McKay (1996). McKay argues that in rave, traveller cultures, eco-protest, and Reclaim the Streets, Britain has contemporary versions of the 1960s 'culture of resistance' (ibid.: 6), although (after Redhead [1993] and Rietveld [1993]) McKay does question the extent to which a political critique is posed by rave. For further discussion of traveller cultures as a form of social movement, see Hetherington (1996; 1998).
4 Reprinted in Goldman (1993: 286; emphasis in original).
5 Reprinted in Goldman (1993: 288).
6 Extract from *Ecstasy*, by Irvine Welsh (1996: 26–7).
7 The final paragraph from the short story called *The Moment* by one of the clubber interviewees, Dionne.

APPENDIX

1 These biographical sketches were accurate at the time of the interviews.

REFERENCES

Aalten, A. (1997) 'Performing the body, creating culture', in K. Davis (ed.) *Embodied Practices: Feminist Perspectives on the Body*, Sage: London, pp. 41–58.

Abel, E.L. (1980) *Marihuana: The First Twelve Thousand Years*, Plenum Press: New York.

Abelson, E.S. (1989) *When Ladies Go A-Thieving: Middle Class Shoplifters in the Victorian Department Store*, Oxford University Press: Oxford.

Abercrombie, N., Keat, R. and Whitley, N. (eds) (1994) *The Authority of the Consumer*, Routledge: London.

Acre, A. and Marsden, T. (1993) 'The social construction of international food: A new research agenda', *Economic Geography* 69(3): 93–111.

Adorno, T. (1976) *Introduction to the Sociology of Music* (translated by E.B. Ashton), Seabury Press: New York.

—— (1973) *Philosophy of Modern Music* (translated by Anne G. Mitchell and Wesley V. Bloomster), Sheed & Ward: London.

Ali, T. (1998) 'What's the matter with our dumbed-down youth?', in *Independent*, 16 February, p. 15.

Ali, T. and Watkins, S. (1998) *Marching in the Streets*, Bloomsbury: London.

Amirou, R. (1989) 'Sociability/"Sociality"', *Current Sociology* 37(1), spring, pp. 115–20 (special issue on the Sociology of Everyday Life).

Aries, P. (1962) *Centuries of Childhood*, Free Press: New York.

Backman, E.L. (1952) *Religious Dances in the Christian Church and in Popular Medicine* (translated by E. Classen), Allen & Unwin: London.

Balding, J. (1997) *Young people in 1996*, University of Exeter: Exeter.

Barnes, R. (1980) *Mods!*, Eel Pie Publishing: London.

Bat (1997) 'Disposable theory: Jungle, metalheadz and pirate radio', unpublished paper, March.

Baudelaire, C. (1964) [1863] *The Painter of Modern Life and Other Essays* (translated and with an introduction by Jonathan Mayne), Phaidon Press: London.

Bauman, Z. (1995a) *Life in Fragments: Essays in Postmodern Morality*, Blackwell: Oxford.

—— (1995b) 'Forms of togetherness', in *Life in Fragments: Essays in Postmodern Morality*, Blackwell: Oxford: pp. 44–71.

—— (1993) *Modernity and Ambivalence*, Polity: Cambridge.

—— (1992) *Intimations of Postmodernity*, Routledge: London.

—— (1988) *Freedom*, Open University Press: Milton Keynes.

Becker, H. (1983) *Outsiders: Studies in the Sociology of Deviance*, Free Press: New York.

Bell, D., Binnie, J., Cream, J. and Valentine, G. (1994) 'All hyped up and no place to go', *Gender, Place and Culture* 1(1): 31–47.

Bellos, A. (1997) 'Media', in N. Saunders (ed.) *Ecstasy Reconsidered*, with a bibliography by Alexander Shulgin, Neal's Yard Press: London, pp. 28–35.

Benjamin, W. (1973) *Charles Baudelaire: A Lyric Poet in the Era of High Capitalism* (translated by H. Zohn), New Left Books: London.

Berridge, V. and Edwards, G. (1981) *Opium and the People: Opiate Use in Nineteenth Century England*, Allen Lane: London.

Blacking, J. (1987) *A Commonsense View of All Music*, Cambridge University Press: Cambridge.

—— (1975) 'Movement, dance, music and the Venda Girls' initiation cycle', in P. Spencer (ed.) *Society and the Dance: The Social Anthropology of Process and Performance*, Cambridge University Press: Cambridge.

—— (1973) *How Musical Is Man?*, University of Washington Press: London.

Blake, W. (1969) *The Complete Writings of Blake: With Variant Readings* (edited by Geoffrey Keynes), Oxford University Press: Oxford.

Blomley, N. (1996) ' "I'd like to dress her all over": Masculinity, power and retail space', in N. Wrigley and M. Lowe (eds) *Retailing, Consumption and Capital: Towards the New Retail Geography*, Longman: Harlow, Essex, pp. 238–56.

Boas, F. (ed.) (1972) [1924] *The Function of Dance in Human Society*, Dance Horizons: New York.

Bocock, R. (1993) *Consumption*, Routledge: London.

Boden, D. and Moltoch, H.L. (1994) 'The compulsion of proximity', in R. Friedland and D. Boden (eds) *NowHere: Space, Time and Modernity*, University of California Press: London, pp. 257–86.

Boethius, U. (1995) 'Youth, the media and moral panics', in J. Fornäs and G. Bolin (eds) *Youth Cultures in Late Modernity*, Sage: London.

Bourdieu, P. (1984) [1979] *Distinction: A Social Critique of the Judgement of Taste* (translated by Richard Nice), Routledge & Kegan Paul: London.

Braidotti, R. (1994) *Nomadic Subjects: Embodiment and Sexual Difference in Contemporary Feminist Theory*, Columbia University Press: New York.

Brinson, P. (1985) 'Epilogue: Anthropology and the study of dance', in P. Spencer (ed.) *Society and the Dance: The Social Anthropology of Process and Performance*, Cambridge University Press: Cambridge, pp. 206–14.

British Crime Survey (1994), HMSO: London.

Brown, A. (1997) 'Let's all have a disco? Football, popular music and democratization', in S. Redhead (ed.) *The Clubcultures Reader: Readings in Popular Cultural Studies*, Blackwell: Oxford, pp. 79–101.

Browning, B. (1995) *Samba: Resistance in Motion*, Indiana University Press: Indianapolis, Indiana.

Burke, T. (1941) *English Night-Life: From Norman Curfew to Present Black-Out*, B.T. Batsford Ltd.: London.

Butler, J. (1993) *Bodies that Matter: On the Discursive Limits of Sex*, Routledge: London.

—— (1990a) *Gender Trouble: Feminism and the Subversion of Identity*, Routledge: London.

—— (1990b) 'Performative acts and gender constitution', in S.-E. Case (ed.) *Performing Feminisms: Feminist Critical Theory and Theatre*, John Hopkins University Press: Baltimore, pp. 270–83.

Byron, G. (1816) *Childe Harold's Pilgrimage*, John Murray: London.

Campbell, C. (1995) 'The Sociology of Consumption', in D. Miller (ed.) *Acknowledging Consumption: A Review of New Studies*, Routledge: London, pp. 96–126.

—— (1987) *The Romantic Ethic and the Spirit of Modern Consumerism*, Blackwell: Oxford.

Campbell, D. (1997) 'Nearly all ravers "have tried drugs" ', *Guardian*, August 6, p. 8.

Canetti, E. (1973) [1960] *Crowds and Power*, translated by Carol Stewart, Penguin: London.

Canon, T. (1993) 'World music', paper presented at the Annual Conference of the Institute of British Geographers, Royal Holloway and Bedford New College, January.

Chambers, I. (1994) *Migrancy, Culture, Identity*, Routledge: London.

—— (1990) *Border Dialogues: Journeys in Postmodernity*, Routledge: London.

—— (1986) *Popular Culture: The Metropolitan Experience*, Methuen: London.

—— (1985) *Urban Rhythms: Pop Music and Popular Culture*, Macmillan: London.

Chaney, D. (1996) *Lifestyles*, Routledge: London.

—— (1990) 'Subtopia in Gateshead: The Metro Centre as cultural form, *Theory, Culture & Society* 7: 49–68.

Chow, R. (1994) *Writing Diaspora: Tactics of Intervention in Contemporary Cultural Studies*, Indiana University Press: Bloomington, Indiana.

—— (1993) 'Listening otherwise, music miniaturized: A different type of revolution', in S. During (ed.) *The Cultural Studies Reader*, Routledge: London, pp. 382–99.

Clarke, D.B. (1996) 'The limits to retail capital', in N. Wrigley and M. Lowe (eds) *Retailing, Consumption and Capital: Towards the New Retail Geography*, Longman: Harlow, Essex, pp. 284–301.

Clarke, J., Hall, S., Jefferson, T. and Roberts, B. (1993) 'Subcultures, cultures and class', in *Resistance Through Rituals: Youth Subcultures in Post-War Britain*, Routledge: London.

Cohen, P. (1980) *Folk Devils and Moral Panics: The Creation of the Mods and Rockers* (2nd edn), Martin Robertson: London.

Cohen, S. (1995) 'Sounding out the city: Music and the sensuous production of place', *Transactions of the Institute of British Geographers* 20(4): 434–46.

—— (1991a) *Rock Culture in Liverpool*, The Clarendon Press: London.

—— (1991b) 'Popular music and urban regeneration: The music industries of Merseyside, *Cultural Studies* 5: 332–46.

Cohn, M. (1992) *Dope Girls*, Lawrence & Wishart: London.

Collin, M. (1997) *Altered State: The Story of Ecstasy Culture and Acid House*, Serpent's Tail: London.

—— (1996) 'Saturday night fever', *The Face* 2(94), July, pp. 110–12.

Cook, I. (1995) 'Constructing the exotic: The case of tropical fruit', in J. Allen and D. Massey (eds) *Geographical Worlds*, Oxford University Press: Oxford, pp. 137–42.

Cook, I. and Crang, P. (1996) 'The world on a plate: Culinary culture, displacement and geographical knowledges', *Journal of Material Culture* 1(1): 131–53.

Cook, M. (1997) 'How to dance in today's modern "club" style', in *MixMag* 2(73), June, pp. 116–17.

Crang, P. and Malbon, B. (1996) 'Consuming geographies: A review essay, *Transactions of the Institute of British Geographers* 21(4): 704–19.

Cranston, M. (1983) *Jean-Jacques*, Allen Lane: London.

Crawford, M. (1992) 'The world in a shopping mall', in M. Sorkin (ed.) *Variations on a Theme Park*, Hill & Wang: New York, pp. 3–30.

Cream, J. (1995) 'Re-solving riddles: The sexed body', in D. Bell and G. Valentine (eds) *Mapping Desire: Geographies of Sexualities*, Routledge: London, pp. 31–40.

Cresswell, T. (1996) *In Place/Out of Place: Geography, Ideology, and Transgression*, University of Minnesota Press: London.

Crewe, L. and Gregson, N. (1998) 'Tales of the unexpected: Exploring car boot sales as marginal spaces of contemporary consumption', *Transactions of the Institute of British Geographers* 23(1): 39–53.

Crewe, L. and Lowe, M. (1995) 'Gap on the map? Towards a geography of consumption and identity', *Environment and Planning A* 27: 1877–98.

Crossley, N. (1995) 'Body techniques, agency and intercorporeality: On Goffman's "Relations in Public"', *Sociology* 29(1): 133–49.

Csikszentmihalyi, M. (1975a) *Beyond Boredom and Anxiety: The Experience of Play in Work and Games*, Jossey-Bass: San Francisco.

—— (1975b) 'Measuring the flow experience in rock dancing', in *Beyond Boredom and Anxiety: The Experience of Play in Work and Games*, Jossey-Bass: San Francisco, pp. 102–122.

—— (1975c) 'Politics of enjoyment', in *Beyond Boredom and Anxiety: The Experience of Play in Work and Games*, Jossey-Bass: San Francisco, pp. 179–206.

Csikszentmihalyi, M. and Rathunde, K. (1993) 'The measurement of flow in everyday life – towards a theory of emergent motivation', *Nebraska Symposium on Motivation* 40: 57–97.

Daily Star (1997) 'Dance of death: New disco craze menaces Britain', front-page editorial, 29 October.

de Beauvoir, S. (1972) *The Second Sex*, Penguin: Harmondsworth.

de Cauter, L. (1994) 'The panoramic ecstasy: On world exhibitions and the disintegration of experience', *Theory, Culture and Society* 10(4): 1–24.

de Certeau, M. (1993) 'Walking in the city', in S. During (ed.) *The Cultural Studies Reader*, Routledge: London, pp. 151–60.

—— (1984) [1974] *The Practice of Everyday Life* (translated by Stephen F. Rendall), University of California Press, California.

—— (1980) 'On the oppositional practices of everyday life', *Social Text* 3: 3–43.

Desenberg, J.T. (1997) *Ecstasy and the Status Quo: Transcending the Invisible Barrier*, electronically published at: http://www.cia.com.au/peril/youth/jesse1.htm.

Desmond, J.C. (ed.) (1997) *Meaning in Motion: New Cultural Studies of Dance*, Duke University Press: London.

de Swaan, A. (1995) 'Widening circles of identification: Emotional concerns in socio-genetic perspective', *Theory, Culture and Society* 12: 25–39.

Dowling, R. (1991) 'Shopping and the construction of femininity in the Woodward's Department Store, Vancouver, 1945 to 1960', MA thesis, Department of Geography, University of British Columbia.

Druglink (1997) 'Rash of surveys confirm that two in five teenagers take drugs across the UK', *Druglink: Newsletter of the ISDD*, May/June, p. 6.

du Gay, P., Hall, S., Jones, L., Mackay, H. and Negus, K. (1997) *Doing Cultural Studies: The Story of the Sony Walkman*, Sage: London.

During, S. (ed.) (1993) *The Cultural Studies Reader*, Routledge: London.

Durkheim, E. (1912) *The Elementary Forms of Religious Life* (translated by J.W. Swain), Allen & Unwin: London.

Dwyer, C. (1998) 'Contested identities: Challenging dominant representations of young British Muslim women', in T. Skelton and G. Valentine (eds) *Cool Places: Geographies of Youth Cultures*, Routledge: London, pp. 50–65.

Dyer, R. (1993) 'Entertainment and Utopia', in S. During (ed.) *The Cultural Studies Reader*, Routledge: London, pp. 271–83.

Edgell, S., Hetherington, K. and Warde, A. (eds) (1997) *Consumption Matters: The Production and Experience of Consumption*, Blackwell: Oxford.

Evans, D. (1997) 'Michel Maffesoli's sociology of modernity and postmodernity: An introduction and critical assessment', *Sociological Review* 45(2): 220–43.

Featherstone, M. (1995) *Undoing Culture: Globalization, Postmodernism and Identity*, Sage: London.

—— (1991) *Consumer Culture and Postmodernism*, Sage: London.

Finkelstein, J. (1991) *The Fashioned Self*, Polity: Cambridge.

Finnegan, R. (1989) *The Hidden Musicans: Making Music in an English Town*, Cambridge University Press, Cambridge.

Flinn, C. (1992) *Strains of Utopia*, Princeton University Press: New Jersey.

Fornäs, J. (1997) 'Text and music revisited', *Theory, Culture and Society* 14(3): 109–23.

Forster, E.M. (1946) [1910] *Howard's End*, Penguin: London.

Foster, S.L. (ed.) (1996) *Corporealities: Dancing Knowledge, Culture and Power*, Routledge: London.

Freud, S. (1961) *Civilization and its Discontents* (translated and edited by James Strachey *et al.*), Hogarth Press: London (standard edn, vol. XXI).

Friedman, J. (ed.) (1994) *Consumption and Identity*, Harwood Academic: Chur, Switzerland.

—— (1992) 'Narcissism, roots and postmodernity: The constitution of selfhood in the global crisis', in S. Lash and J. Friedman (eds) *Modernity and Identity*, Blackwell: Oxford, pp. 331–66

Frith, S. (1996) *Performing Rites: On the Value of Popular Music*, Oxford University Press: Oxford.

—— (1995) 'Editorial', *New Formations* 27, winter 1995/6, pp. v–xii (special issue on Performance Matters).

—— (1992) 'The cultural study of popular music', in L. Grossberg *et al.* (eds) *Cultural Studies*, Routledge: London, pp. 174–86.

—— (1988) *Music for Pleasure*, Polity: Cambridge.

—— (1987) 'Copyright and the music business', *Popular Music* 7: 57–75.

—— (1978) *The Sociology of Rock*, Constable: London.

Garratt, S. (1998) *Adventures in Wonderland: A Decade of Club Culture*, Headline: London.

Gelder, K. (1997) 'Introduction to Part 2', in K. Gelder and S. Thornton (eds) *The Subcultures Reader*, Routledge: London, pp. 83–9.

Gelder, K. and Thornton, S. (eds) (1997) *The Subcultures Reader*, Routledge: London.

Gendron, B. (1986) 'Theodor Adorno Meets the Cadillacs', in T. Modleski (ed.) *Studies in Entertainment: Critical Approaches to Mass Culture*, Indiana University Press: Bloomington, Indiana, pp. 18–36.

Gibson, C. and Zagora, D. (1997) *Rave Culture in Sydney: Mapping the Social Construction of Youth Spaces*, unpublished paper, Rave Research Collective, Department of Geography, University of Sydney, Australia.

Giddens, A. (1964) 'Notes on the concepts of play and leisure', *Sociological Review* 12(1).

Gilbert, J. (1997) 'Soundtrack to an uncivil society: Rave culture, the Criminal Justice Act and the politics of modernity, *New Formations* 31: 5–22.

Glennie, P. and Thrift, N. (1996) 'Consumption, shopping, gender', in N. Wrigley and M. Lowe (eds) *Retailing, Consumption and Capital: Towards the New Retail Geography*, Longman: Harlow, Essex, pp. 221–37.

—— (1992) 'Modernity, urbanism and modern consumption, in *Environment and Planning D: Society and Space* 10: 423–43.

Goffman, E. (1971) *Relations in Public: Microstudies of the Public Order*, Penguin: London.

—— (1969) *Strategic Interaction*, University of Pennsylvania Press: Philadelphia.

—— (1968) *Stigma: Notes on the Management of Spoiled Identity*, Penguin: Harmondsworth, Middlesex.

—— (1967) *Interaction Ritual: Essays on Face-to-Face Behavior*, Penguin: London.

—— (1963) *Behavior in Public Places: Notes on the Social Organization of Gatherings*, Free Press: New York.

—— (1961) *Encounters: Two Studies in the Sociology of Interaction*, Bobbs-Merrill: Indianapolis, Indiana.

—— (1959) *The Presentation of Self in Everyday Life*, Penguin: London.

Goldman, A. (1993) *Sound Bites*, Abacus: London.

—— (1978) *Disco*, Abacus: London.

Goldman, R. (1992) *Reading Ads Socially*, Routledge: London.

Goodman, J., Lovejoy, P.E. and Sherratt, A. (eds) (1995) *Consuming Habits: Drugs in History and Anthropology*, Routledge: London.

Gore, G. (1997) 'The beat goes on: Trance, dance and tribalism in rave culture', in H. Thomas (ed.) *Dance in the City*, Macmillan: London, pp. 50–67.

Goss, J. (1993) 'The "magic of the mall": An analysis of form, function and meaning in the contemporary retail built environment', *Annals of the Association of American Geographers* 83: 18–47.

—— (1992) 'Modernity and post modernity in the retail landscape', in K. Anderson and J. Gale (eds) *Inventing Places: Studies in Cultural Geography*, Longman Cheshire: Melbourne.

Gotfrit, L. (1988) 'Women dancing back: Disruption and the politics of pleasure', *Journal of Education* 170(3).

Gowan, J.C. (1975) *Trance, Art, and Creativity*, privately printed: Northridge, California.

Greenhalgh, P. (1988) *Ephemeral Visions: The 'Exposition Universelles', Great Exhibitions and World's Fairs 1851–1939*, Manchester University Press: Manchester.

Gregson, N. and Crewe, L. (1997a) 'Performance and possession: Rethinking the act of purchase in the light of the car boot sale', *Journal of Material Culture* 2(2): 241–63.

—— (1997b) 'The bargain, the knowledge and the spectacle: Making sense of consumption in the space of the car boot sale', *Environment and Planning D: Society and Space* 15(1): 87–112.

—— (1994) 'Beyond the high street and the mall: Car boot fairs and the new geographies of consumption in the 1990s', *Area* 26(3): 261–67.

Gregson, N., Crewe, L. and Longstaff, B. (1997) 'Excluded spaces of regulation: Car boot sales as an enterprise culture out of control?', *Environment and Planning A* 29(10): 1717–37.

Griffiths, V. (1988) 'Stepping out: The importance of dancing for young women', in E. Wimbush and M. Talbot (eds) *Relative Freedoms: Women and Leisure*, Open University Press: Milton Keynes.

Grossberg, L. (1984) 'Another boring day in paradise', in *Popular Music* 4: 225–60.

Haden-Guest, A. (1997) *'The Last Party': Studio 54, Disco, and the Culture of the Night*, William Morrow and Company, Inc.: New York.

Halfacree, K.H. (1996) 'Out of place in the country: Travelers and the "rural idyll" ', *Antipode* 28: 42–71.

Halfacree, K.H. and Kitchin, R.B. (1996) ' "Madchester rave on": Placing the fragments of popular music', *Area* 28(1): 47–55.

Hall, E.T. (1955) 'The anthropology of manners', *Scientific American*, April, pp. 84–90.

Hall, S. (1996) 'Introduction: Who needs "identity"?', in S. Hall and P. du Gay (eds) *Questions of Cultural Identity*, Sage: London, pp. 1–17.

—— (1992) 'The question of cultural identity', in S. Hall, D. Held and T. McGrew (eds) *Modernity and its Futures*, Polity: Cambridge, pp. 274–325.

—— (1991) 'Old and new identities, old and new ethnicities', in A.D. King (ed.) *Culture, Globalization and the World System: Contemporary Conditions for the Representation of Identity*, Macmillan: London, pp. 41–68.

—— (1984) 'Reconstruction work', *Ten.8* 16: 2–9.

Hall, S. and du Gay, P. (eds) (1996) *Questions of Cultural Identity*, Sage: London.

Hall, S. and Jefferson, T. (eds) (1993) *Resistance Through Rituals: Youth Subcultures in Post-War Britain*, Routledge: London (first published in 1975 as Working Papers in Cultural Studies 7/8).

Hanna, J.L. (1987) [1979] *To Dance is Human: A Theory of Nonverbal Communication* (2nd edn), University of Chicago Press: Chicago.

Happold, F.C. (1981) [1963] *Mysticism: A Study and an Anthology* (revised edn), Penguin: Harmondsworth.

Haraway, D.J. (1991) *Simians, Cyborgs and Women: The Reinvention of Nature*, Free Association Books: London.

Hebdige, D. (1988) *Hiding in the Light: On Images and Things*, Routledge: London.

—— (1983) 'Posing...threats, striking...poses: Youth, surveillance and display', *SubStance* 37/38, reprinted in K. Gelder and S. Thornton (eds) *The Subcultures Reader*, Routledge: London, pp. 393–405.

—— (1979) *Subculture: The Meaning of Style*, Methuen: London.

Hecker, J.F. (1970) *The Dancing Mania in the Middle Ages*, Chapman & Hall: New York.

Hediger, H. (1955) *Studies of the Psychology and Behavior of Captive Animals in Zoos and Circuses*, Butterworth Scientific Publications: London.

Hemment, D. (1997) 'e is for *Ekstasis*', *New Formations* 31: 23–38.

Henderson, S. (1993) *Women, Sexuality and Ecstasy Use: The Final Report*, paper published by Lifeline, a non-statutory drug agency in Manchester (101 Oldham Street, Manchester M4 1LW).

Henry, J. (1992) 'Ecstasy and dance of death', *British Medical Journal* 305(4), July, pp. 5–6.

Hetherington, K. (1997) *The Badlands of Modernity: Heterotopia and Social Ordering*, Routledge: London.

—— (1996) 'Identity formation, space and social centrality', in *Theory, Culture and Society* 13(4): 33–52.

—— (1994) 'The contemporary significance of Schmalenbach's concept of the *Bund*', *Sociological Review* 42(1): 1–25.

—— (1992) 'Stonehenge and its festival: Spaces of consumption, in R. Shields (ed.) *Lifestyle Shopping: The Subject of Consumption*, Routledge: London, pp. 83–98.

Hollows, J. and Milestone, K. (1998) 'Welcome to Dreamsville: A history and geography of northern soul', in A. Leyshon, D. Matless and G. Revill (eds) *The Place of Music*, The Guildford Press: New York, pp. 83–103.

Hopkins, J. (1991) 'West Edmonton Mall as a centre for social interaction', *Canadian Geographer* 35: 268–79.

—— (1990) 'West Edmonton Mall: Landscapes of myths and elsewhereness', *Canadian Geographer* 34: 2–17.

Hornby, N. (1995) *High Fidelity*, Indigo: London.

Hudson, R. (1995) 'Making music work? Alternative regeneration strategies in a deindustrialized locality: The case of Derwentside', *Transactions of the Institute of British Geographers* 20(4): 460–73.

Hughes, W. (1994) 'In the empire of the beat: Discipline and disco', in A. Ross and T. Rose (eds) *Microphone Fiends: Youth Music and Youth Culture*, Routledge: London, pp. 147–57.

Huizinga, J. (1969) *Homo Ludens: A Study of the Play Element in Culture*, Paladin: London.

Huxley, A. (1972) *Island*, Perennial Library: London.

—— (1961) *The Doors of Perception*, Penguin: London.

Inglis, B. (1989) *Trance: A Natural History of Altered States of Mind*, Grafton: London.

ISDD (1996) *Drug Notes 8: Ecstasy*, ISDD (Institute for the Study of Drug Dependence): London.

Jackson, E. and Johnson, D. (1991) 'Geographic implications of mega-malls, with specific reference to West Edmonton Mall', in *Canadian Geographer* 35: 226–31.

Jackson, M. (1989) *Paths toward a Clearing*, Indiana University Press: Bloomington, Indiana.

Jackson, P. (1995) 'Editorial: Changing geographies of consumption', *Environment and Planning A* 27: 1875–6.

—— (1989) *Maps of Meaning*, Unwin Hyman: London.

—— (1988) 'Street life: The politics of Carnival', *Environment and Planning D: Society and Space* 6: 213–27.

Jackson, P. and Thrift, N. (1995) 'Geographies of consumption', in D. Miller (ed.) *Acknowledging Consumption: A Review of New Studies*, Routledge: London, pp. 204–37.

James, M. (1997) *State of Bass – Jungle: The Story So Far*, Boxtree: London.

James, W. (1961) [1892] *Psychology: The Briefer Course*, Harper & Row: New York.

Jordan, T. (1995) 'Collective bodies: Raving and the politics of Gilles Deleuze and Felix Guattari', *Body and Society* 1(1):125–44.

Jung, C.G. (1973) [1963] *Memories, Dreams, Reflections*, Routledge & Kegan Paul: London.

—— (1957) *The Collected Works of C.G. Jung: Volume 1 – Psychiatric Studies* (edited by Sir Herbert Reed and translated by R.F.C. Hull), Routledge & Kegan Paul: London.

Keith, M. and Pile, S. (eds) (1993) *Place and the Politics of Identity*, Routledge: London.

Kellner, D. (1983) 'Critical theory, commodities and the consumer society, *Theory, Culture and Society* 1(3): 66–83 (special issue on Consumer Culture).

Kendon, A. (1988) 'Goffman's approach to face-to-face interaction', in P. Drew and A. Wootton (eds) *Erving Goffman: Exploring the Interaction Order*, Polity: Cambridge, pp. 14–40.

Kong, L. (1995a) 'Music and cultural politics: Ideology and resistance in Singapore', *Transactions of the Institute of British Geographers* 20(4): 447–59.

—— (1995b) 'Popular music in geographical analyses', *Progress in Human Geography* 19: 183–98.

Kossoff, J. (1995) 'Party politics', *Time Out*, November 29–December 6, pp. 12–13.

Kowinski, W. (1985) *The Malling of America: An Inside Look at the Consumer Paradise*, William Morrow: New York.

Kruse, H. (1993) 'Subcultural identity in alternative music', *Popular Music* 12: 33–41.

Kureishi, H. and Savage, J. (eds) (1995) *The Faber Book of Pop*, Faber & Faber: London.

Lash, S. and Urry, J. (1994) *Economies of Signs and Space*, Sage: London.

Laski, M. (1980) *Everyday Ecstasy*, Thames & Hudson: London.

—— (1961) *Ecstasy: A Study of some Secular and Religious Experiences*, The Cresset Press: London.

Law, J. (1997) 'Dancing on the bar: Sex, money and the uneasy politics of the third space', in S. Pile and M. Keith (eds) *Geographies of Resistance*, Routledge: London, pp. 105–23.

—— (1994) *Organizing Modernity*, Blackwell: Oxford.

Leary, T. (1973) *The Politics of Ecstasy*, Granada Publishing: London.

—— (1968) *High Priest*, World Publishing Company: Cleveland.

Le Bon, G. (1930) [1896] *The Crowd: A Study of the Popular Mind*, Ernest Benn Ltd.: London.

Lee, M. (1993) *Consumer Culture Re-born: The Cultural Politics of Consumption*, Routledge: London.

Lemert, C. and Branaman, A. (eds) (1997) *The Goffman Reader*, Blackwell: Oxford.

Leppert, R. and McClary, S. (eds) (1987) *Music and Society: The Politics of Composition, Performance and Reception*, Cambridge University Press: Cambridge.

Levine, J. (1987) 'Contra dance in New York: Longways for as many as will', in G.A. Fine (ed.) *Meaningful Play, Playful Meaning*, Human Kinetics Publishers, University of Minnesota, Illinois, pp. 193–205.

Lewis, I.M. (1989) *Ecstatic Religion: A Study of Shamanism and Spirit Possession* (2nd edn), Routledge: London.

Ley, D. and Olds, K. (1988) 'Landscape as spectacle: World's fairs and the culture of heroic consumption', *Environment and Planning D: Society and Space* 6: 191–212.

Leyshon, A., Matless, D. and Revill, G. (eds) (1998) *The Place of Music*, The Guildford Press: New York

Lofland, L. (1973) *A World of Strangers: Order and Action in Urban Public Space*, Basic Books: New York.

Longhurst, R. (1997) '(Dis)embodied geographies', *Progress in Human Geography* 21(4): 486–501.

—— (1995) 'The body and geography', *Gender, Place and Culture* 2(1): 97–105.

Lovatt, A. (1996) 'The city goes soft: Regeneration and the night-time economy', in D. Wynne and J. O'Connor (eds) *From the Margins to the Centre*, Arena: Aldershot, pp. 141–168.

Ludwig, A.M. (1969) 'Altered states of consciousness', in C.T. Tart (ed.) *Altered States of Consciousness: A Book of Readings*, John Wiley & Sons, Inc: New York, pp. 9–22.

Lunt, P. and Livingstone, S. (1992) *Mass Consumption and Personal Identity: Everyday Economic Experience*, Open University Press: Milton Keynes.

Lury, C. and Warde, A. (1997) 'Investments in the imaginary consumer: Conjectures regarding power, knowledge and advertising', in M. Nava *et al.* (eds) *Buy This Book: Studies in Advertising and Consumption*, Routledge: London, pp. 87–102.

McCracken, G. (1988) *Culture and Consumption: New Approaches to the Symbolic Character of Goods and Activities*, Indiana University Press: Bloomington, Indiana.

McKay, G. (1996) *Senseless Acts of Beauty: Cultures of Resistance since the Sixties*, Verso: London.

McRobbie, A. (1994a) *Postmodernism and Popular Culture*, Routledge: London.

—— (1994b) 'Shut up and dance: Youth culture and changing modes of femininity', in A. McRobbie, *Postmodernism and Popular Culture*, Routledge: London, pp. 155–76.

—— (1994c) 'Different, youthful subjectivities: Towards a cultural sociology of youth, in *Postmodernism and Popular Culture*, Routledge: London, pp. 177–97.

—— (1991) *Feminism and Youth Culture: From 'Jackie' to 'Just Seventeen'*, Macmillan: London.

Madge, C., Raghuram, P., Skelton, T., Willis, K. and Williams, J. (1997) 'Methods and methodologies in feminist geographies: Politics, practice and power', in Women and Geography Study Group, *Feminist Geographies: Explorations in Diversity and Difference*, Longman: London.

Maffesoli, M. (1997) 'The return of Dionysus', in P. Sulkunen, J. Holmwood, H. Radner and G. Schulze (eds) *Constructing the New Consumer Society*, Macmillan: London, pp. 21–37.

—— (1996a) [1985] *Ordinary Knowledge: An Introduction to Interpretative Sociology* (translated by David Macey), Polity: Cambridge.

—— (1996b) [1993] *The Contemplation of the World: Figures of Community Style* (translated by Susan Emmanuel), University of Minnesota Press: Minneapolis.

—— (1995) [1988] *The Time of the Tribes: The Decline of Individualism in Mass Society* (translated by Don Smith), Sage: London.

—— (1993a) [1985] *The Shadow of Dionysus: A Contribution to the Sociology of the Orgy* (translated by Cindy Linse and Mary Kristina Palmquist), SUNY Press: New York.

—— (1993b) *La Contemplation du monde: Figures du style communautaire*, Editions Grasset & Fasquelle: Paris.

—— (1989) 'The sociology of everyday life (epistemological elements)', *Current Sociology* 37(1), spring, pp. 1–16 (special issue on the Sociology of Everyday Life).

—— (1988a) *Le Temps des tribus*, Méridiens Klincksieck: Paris.

—— (1988b) 'Jeux de Masques: Postmodern Tribalism', *Design Issues* IV(1&2): 141–51 (special issue).

—— (1987) 'Sociality as legitimation of sociological method' (translated by Milli Kerkham), *Current Sociology* 35(2): 69–87.

—— (1985a) *L'Ombre de Dionysos: Contribution à une sociologie de l'orgie*, Méridiens Klincksieck: Paris.

—— (1985b) *La Connaissance ordinaire: Précis de sociologie compréhensive*, Méridiens Klincksieck: Paris.

Mahtani, M.K. (1998) 'Mapping the paradoxes of multiethnicity: Stories of multiethnic women in Toronto, Canada', unpublished Ph.D. thesis, Department of Geography, University College London, 26 Bedford Way, London WC1H 0AP.

Malbon, B. (1998) 'Ecstatic geographies: Clubbing, crowds and playful vitality', unpublished Ph.D. thesis, Department of Geography, University College London, 26 Bedford Way, London WC1H 0AP.

Malcolmson, S.L. (1995) 'Disco dancing in Bulgaria', in M. Henderson (ed.) *Borders, Boundaries and Frames: Cultural Criticism and Cultural Studies*, Routledge: London, pp. 133–41.

Malm, K. and Wallis, R. (1993) *Media Policy and Music Activity*, Routledge: London.

Marcuse, H. (1970) 'Art as a form of reality', in *On the Future of Art*, Viking Press: New York, pp. 123–34.

Marcus, G. (1992) 'Past, present and emergent identities: Requirements for ethnographies of late twentieth-century modernity worldwide', in S. Lash and J. Friedman (eds) *Modernity and Identity*, Blackwell: Oxford, pp. 309–30.

Mauss, M. (1979) [1950] 'Body techniques', in *Sociology and Psychology: Essays* (translated by Ben Brewster), Routledge & Kegan Paul: London, pp. 97–123.

Mennell, S., Murcott, A. and van Otterloo, A. (1992) *The Sociology of Food: Eating, Diet and Culture*, Sage: London.

Menuhin, Y. (1972) *Theme and Variations*, Heinemann Educational: London.

Merchant, J. and MacDonald, R. (1994) 'Youth and rave culture: Ecstasy and health', in *Youth and Policy: The Journal of Critical Analysis* 45, summer, pp. 16–38.

Merleau-Ponty, M. (1962) *Phenomenology of Perception*, Routledge & Kegan Paul: London.

Miller, D. (ed.) (1995) *Acknowledging Consumption: A Review of New Studies*, Routledge: London.

—— (1989) *Material Culture and Mass Consumption*, Blackwell: Oxford.

Mintel (1994) *Nightclubs and Discotheques*, Market Intelligence International Group Ltd.: London.

—— (1996) *Nightclubs and Discotheques*, Market Intelligence International Group Ltd.: London.

Moneta, G.B. and Csikszentmihalyi, M. (1996) 'The effect of perceived challenges and skills on the quality of subjective experience', *Journal of Personality* 64(2): 275–310.

More, T. (1965) [1516] *Utopia*, Penguin: London.

Morris, M. (1988) 'Things to do with shopping centres', in S. Sheridan (ed.) *Grafts: Feminist Cultural Criticism*, Verso: London, pp. 193–225.

Mort, F. (1989) 'The politics of consumption', in S. Hall and M. Jacques (eds) *New Times: The Changing Face of Politics in the 1980s*, Lawrence & Wishart: London, pp. 160–72.

Mullan, R. *et al.* (1997) *Young People's Drug use at Rave/Dance Events*, Crew 2000: Edinburgh.

Nasmyth, P. (1985) 'MDMA we're all crazy now: Ecstasy's arrival in Britain, in R. Benson (ed.) (1997) *NIGHTFEVER: Club Writing in* The Face *1980–1997*, Boxtree: London, pp. 74–8.

Nava, M. (1992) *Changing Cultures: Feminism, Youth and Consumerism*, Sage: London.

Nava, M., Blake, A., MacRury, I. and Richards, B. (1997) *Buy This Book: Studies in Advertising and Consumption*, Routledge: London.

Norris, C. (ed.) (1989) *Music and the Politics of Culture*, Lawrence & Wishart: London.

O'Connor, B. (1997) 'Safe sets: Women, dance and 'communitas', in H. Thomas (ed.) *Dance In The City*, Macmillan: London, pp. 149–72.

O'Connor, J. and Wynne, D. (1995) 'The city and the night-time economy', *Planning, Practice and Research* 10(2).

—— (1994) *From the Margins to the Centre: Cultural Production and Consumption in the Post-Industrial City*, Working Papers in Cultural Studies 7, Manchester Institute for Popular Culture: Manchester.

Perri 6 (1997) 'Dangerous hallucination', *Guardian*, Society section, 5 November, p. 6

Peterson, R.A. and Simkus, A. (1992) 'How musical tastes mark occupational status groups', in M. Lamont and M. Fournier (eds) *Cultivating Differences: Symbolic Boundaries and the Making of Inequality*, University of Chicago Press: London, pp. 152–86.

Pile, S. (1997) 'Introduction: Opposition, political identities and spaces of resistance', in S. Pile and M. Keith (eds) *Geographies of Resistance*, Routledge: London, pp. 1–32.

Pile, S. and Keith, M. (eds) (1997) *Geographies of Resistance*, Routledge: London.

Pini, M. (1997a) 'Cyborgs, nomads and the raving feminine', in H. Thomas (ed.) *Dance in the City*, Macmillan: London, pp. 111–29.

—— (1997b) 'Women and the early British rave scene', in A. McRobbie (ed.) *Back to Reality: New Readings in Cultural Studies*, Manchester University Press: Manchester, pp. 152–69.

Pinnegar, S. (1999) 'Millennial thought into practice at the Earth Centre: A study in the art of translation', unpublished Ph.D. thesis, Department of Geography, University College London, 26 Bedford Way, London WC1H 0AP.

—— (1995) 'Men, masculinities and feminist theory: Having a gender too', unpublished MA thesis, Department of Geography, Carleton University, Ottawa, Ontario.

Polhemus, T. (1993) 'Dance, gender and culture', in H. Thomas (ed.) *Dance, Gender and Culture*, Macmillan: London, pp. 3–15.

Pred, A. (1994) *Recognising European Modernities: A Montage of the Present*, Routledge: London.

Raban, J. (1974) *Soft City*, Harvill: London.

Ramsey, K. (1997) 'Vodou, nationalism, and performance: The staging of folklore in mid-twentieth-century Haiti', in J.C. Desmond (ed.) *Meaning in Motion: New Cultural Studies of Dance*, Duke University Press: London, pp. 345–78.

Redhead, S. (ed.) (1997) *The Clubcultures Reader: Readings in Popular Cultural Studies*, Blackwell: Oxford.

—— (ed.) (1993) *Rave Off: Politics and Deviance in Contemporary Youth Culture*, Avebury: Aldershot.

Regev, M. (1997) 'Rock aesthetics and musics of the world', *Theory, Culture and Society* 14(3): 125–42.

Release (1997) *Release Drugs and Dance Survey: An Insight into the Culture*, Release: London.

Reynolds, S. (1998) *Energy Flash: A Journey Through Rave Music and Dance Culture*, Picador: London.

Richard, B. and Kruger, H.H. (1998) 'Ravers' paradise?: German youth cultures in the 1990s', in T. Skelton and G. Valentine (eds) *Cool Places: Geographies of Youth Cultures*, Routledge: London, pp. 161–74.

Rietveld, H. (1998a) *Pure Bliss: Intertextuality in House Music*, electronically published at: http://www.finearts.yorku.ca/DE/rietveld.html.

—— (1998b) *This is our House*, Arena: London.

—— (1993) 'Living the dream', in S. Redhead (ed.) *Rave Off: Politics and Deviance in Contemporary Youth Culture*, Avebury Press: Aldershot, pp. 41–78.

Rose, G. (1995) 'Geography and gender: Cartographies and corporealities', *Progress in Human Geography* 19(4): 544–8.

Ross, A. (1994) 'Introduction', in A. Ross and T. Rose (eds) *Microphone Fiends: Youth Music and Youth Culture*, Routledge: London, pp. 1–13.

Ross, A. and Rose, T. (eds) (1994) *Microphone Fiends: Youth Music and Youth Culture*, Routledge: London.

Rouget, G. (1985) [1980] *Music and Trance: A Theory of the Relations between Music and Possession* (translated from the French by Brunhilde Biebuyck in collaboration with the author), University of Chicago Press: Chicago.

Royce, A.P. (1977) *The Anthropology of Dance*, Indiana University Press: Bloomington, Indiana.

Sack, R. (1992) *Place, Modernity and the Consumer's World: A Relational Framework for Geographical Analysis*, John Hopkins University Press: Baltimore, Maryland.

Sadgrove, J. (1997) 'Going with the flow', in *Guardian*, G2 section, 2 September, pp. 18–19.

Said, E. (1991) *Musical Elaborations*, Chatto & Windus: London.

Saldanha, A. (1998) *Music, Space, Identity: Global Youth/Local Others in Bangalore, India*, published electronically at: http://www.cia.com.au/peril/youth/arun-msi.htm (or e-mail the author at: jsaldanh@vub.ac.be).

Saunders, N. (ed.) (1997) *Ecstasy Reconsidered*, with a bibliography by Alexander Shulgin, Neal's Yard Press: London.

—— (1996) *How the Media Reports Ecstasy*, published electronically at http://www.obsolete.com/ecstasy.media.html.

—— (ed.) (1995) *Ecstasy and the Dance Culture*, with contributions by Mary Anna Wright and an annotated bibliography by Alexander Shulgin, Neal's Yard Press: London.

—— (ed.) (1994) *E for Ecstasy*, Neal's Yard Press: London.

Schechner, R. (1993) *The Future of Ritual: Writings on Culture and Performance*, Routledge: London.

—— (1981) 'Restoration of behavior', *Studies in Visual Communication* 7: 2–45.

Scheff, T.J. (1990) *Microsociology: Discourse, Emotion, and Social Structure*, University of Chicago Press: Chicago.

Scheler, M. (1954) *The Nature of Sympathy* (translated from the German by Peter Heath), Routledge & Kegan Paul: London.

Schiller, F. (1967) [1795] *On the Aesthetic Education of Man: In a Series of Letters*, Clarendon Press: Oxford.

Schmalenbach, H. (1922) 'Die soziologische Kategorie des Bundes', in *Die Dioskuren, Vol. 1*, München, pp. 35–105.

Schutz, A. (1964) 'Making music together: A study in social relationship', in *Collected Papers, Volume 2: Studies in Social Theory* (edited and with an introduction by Arvid Brodersen), Martinus Nijhoff: The Hague, pp. 159–78.

Scott, J.C. (1990) *Domination and the Arts of Resistance: Hidden Transcripts*, Yale University Press: London.

—— (1985) *Weapons of the Weak: Everyday Forms of Peasant Resistance*, Yale University Press: New Haven.

Sennett, R. (1990) *The Conscience of the Eye: The Design and Social Life of Cities*, Faber & Faber: London.

Shepherd, J. (1991) *Music as Social Text*, Polity: Cambridge.

Shields, R. (1995) 'Foreword: Masses or tribes?, in M. Maffesoli, *The Time of the Tribes: The Decline of Individualism in Mass Society*, Sage: London, pp. ix–xii.

—— (ed.) (1992a) *Lifestyle Shopping: The Subject of Consumption*, Routledge: London.

—— (1992b) 'Spaces for the subject of consumption', in R. Shields (ed.) *Lifestyle Shopping: The Subject of Consumption*, Routledge: London, pp. 1–20.

—— (1992c) 'The individual, consumption cultures and the fate of community', in R. Shields (ed.) *Lifestyle Shopping: The Subject of Consumption*, Routledge: London, pp. 99–113.

—— (1991a) *Places on the Margin: Alternative Geographies of Modernity*, Routledge: London.

—— (1991b) 'Book review essay: Maffesoli, M. (1990) *Au Creux des apparences* [*In the Hollow of Appearances*], Plon: Paris', in *Theory, Culture and Society* 8: 179–83.

—— (1989) 'Social spatialization and the built environment: West Edmonton Mall', *Environment and Planning D: Society and Space* 7: 147–64.

Shilling, C. (1993) *The Body and Social Theory*, Sage: London.

Shivelbusch, W. (1992) [1980] *Tastes of Paradise: A Social History of Spices, Stimulants and Intoxicants*, Pantheon: New York.

Shulgin, A. and Shulgin, A. (1991) *PIHKAL (Phenethylamines I Have Known and Loved): A Chemical Love Story*, Transform Press: Berkeley, California.

Simmel, G. (1997) [1905] 'The philosophy of fashion', in D. Frisby and M. Featherstone (eds) *Simmel On Culture: Selected Writings*, Sage: London, pp. 187–206.

—— (1981) *Sociologie et épistémologie*, Plon: Paris.

—— (1950) [1903] 'The metropolis and mental life, in K.H. Wolff (ed.) *The Sociology of Georg Simmel* (translated, edited and with an introduction by Kurt H. Wolff), Free Press: New York, pp. 409–24.

Skelton, T. and Valentine, G. (eds) (1998) *Cool Places: Geographies of Youth Cultures*, Routledge: London.

Smith, S. (1997) 'Beyond geography's visible worlds: A cultural politics of music', *Progress in Human Geography* 21(4): 502–29.

—— (1994) 'Soundscape', *Area* 26(3): 232–40.

Snyder, A.F. (1974) 'The dance symbol', in T. Comstock (ed.) *New Dimensions in Dance Research: Anthropology and Dance*, Committee on Research in Dance (CORD): New York.

Solowij, N., Hall, W. and Lee, N. (1992) 'Recreational MDMA use in Sydney: A profile of "ecstasy" users and their experiences with the drug', *British Journal of Addiction* 87: 1161–72.

Sommer, R. (1959) 'Studies in personal space', in *Sociometry* 22: 247–60.

Sorkin, M. (ed.) (1992) *Variations on a Theme Park*, Hill & Wang: New York.

Spencer, P. (ed.) (1985) *Society and the Dance: The Social Anthropology of Process and Performance*, Cambridge University Press: Cambridge.

Stevens, J. (1993) [1987] *Storming Heaven: LSD and the American Dream*, Flamingo: London.

Stoller, P. (1989) *The Taste of Ethnographic Things: The Senses in Anthropology*, University of Pennsylvania Press: Philadelphia.

Storr, A. (1992) *Music and the Mind*, HarperCollins: London.

Stravinsky, I. and Craft, R. (1962) *Expositions and Developments*, Faber & Faber: London.

Straw, W. (1993) 'Characterizing rock music culture: The case of heavy metal', in S. During (ed.) *The Cultural Studies Reader*, Routledge: London, pp. 368–81.

—— (1991) 'Systems of articulation, logics of change: Communities and scenes in popular music', *Cultural Studies* 6: 368–88.

Sulkunen, P., Holmwood, J., Radner, H. and Schulze, G. (eds) (1997) *Constructing the New Consumer Society*, Macmillan: London.

Tart, C.T. (ed.) (1969) *Altered States of Consciousness: A Book of Readings*, John Wiley & Sons, Inc.: New York.

Tester, K. (1993) *The Life and Times of Post-Modernity*, Routledge: London.

Thomas, H. (ed.) (1997) *Dance in the City*, Macmillan: London.

—— (1996) 'Dancing the Difference', *Women's Studies International Forum* 19(5): 505–11.

—— (1995) *Dance, Modernity and Culture: Explorations in the Sociology of Dance*, Routledge: London.

—— (ed.) (1993) *Dance, Gender and Culture*, Macmillan: London.

Thornton, S. (1995) *Club Cultures: Music, Media and Subcultural Capital*, Polity: Cambridge.

—— (1994) 'Moral panic, the media and British rave culture', in A. Ross and T. Rose (eds) *Microphone Fiends: Youth Music and Youth Culture*, Routledge: London, pp. 177–92.

Thrift, N. (1997) 'The still point: Resistance, expressive embodiment and dance', in S. Pile and M. Keith (eds) *Geographies of Resistance*, Routledge: London, pp. 124–51.

—— (1996) 'The still point: Resistance, expressive embodiment and dance', paper presented at Geographies of Domination and Resistance conference, University of Glasgow, September.

Tomlinson, A. (ed.) (1990) *Consumption, Identity and Style: Marketing, Meanings and the Packaging of Pleasure*, Routledge: London.

Toop, D. (1995) *Ocean of Sound: Aether Talk, Ambient Sounds and Imaginary Worlds*, Serpent's Tail: London.

—— (1991) *Rap Attack 2: African Rap to Global Hip Hop*, Serpent's Tail: London.

Touraine, A. (1995) *Critique of Modernity*, Blackwell: Oxford.

Travis, A. (1997) 'Drug takers "not all losers"', *Guardian*, 5 November, p. 6.

Turner, V. (1983) 'Body, brain and culture', *Zygon* 18: 221–45.

—— (1982) *From Ritual to Theatre: The Human Seriousness of Play*, Performing Arts Journal Publication: New York.

—— (1974) *Dramas, Fields, Metaphors*, Cornell University Press: Ithaca, New York.

—— (1969) *The Ritual Process*, Routledge & Kegan Paul: London.

University of Exeter (1992) *Young People in 1992*, Schools Health Education Unit, University of Exeter: Exeter.

Urquía, N. (1996) ' "Now hear this!": A study into the roles of British reggae deejays', unpublished M.Sc. dissertation, School of Education, Politics and Social Science, South Bank University, London.

Valentine, G. (1995) 'Creating transgressive space: The music of kd lang', in *Transactions of the Institute of British Geographers* 20(4): 474–85.

—— (1993) 'Negotiating and managing multiple sexual identities: Lesbian time-space strategies', *Transactions of Institute of British Geographers* 18: 237–48.

van Leeuwen, T. (1988) 'Music and ideology: Notes towards a socio-semotics of Mass Media Music, in T. Threadgold (ed.) *Sydney Association for Studies in Society and Culture Working Papers* 2(1): 29–30.

Ward, A. (1997) 'Dancing around meaning (and the meaning around dance)', in H. Thomas (ed.) *Dance in the City*, Macmillan: London, pp. 3–20.

—— (1993) 'Dancing in the dark: Rationalism and the neglect of social dance', in H. Thomas, (ed.) *Dance, Gender and Culture*, Macmillan: London, pp. 16–33.

Warde, A. (1996) *Consumption, Food and Taste*, Sage: London.

—— (1994a) 'Consumption, identity and uncertainty', *Sociology* 28(4): 877–98.

—— (1994b) 'Consumers, identity and belonging: Reflections on some theses of Zygmunt Bauman', in N. Abercrombie, R. Keat and N. Whitley (eds) *The Authority of the Consumer*, Routledge: London, pp. 58–74.

Weber, M. (1976) [1905] *The Protestant Ethic and the Spirit of Capitalism*, George Allen & Unwin: London.

Wells, H.G. (1933) *The Shape of Things to Come*, Heinemann: London.

—— (1906) *In the Days of the Comet*, Heinemann: London.

Welsh, I. (1996) *Ecstasy: Three Tales of Chemical Romance*, Jonathan Cape: London.

—— (1995) *Marabou Stork Nightmares*, Jonathan Cape: London.

—— (1993) *Trainspotting*, Jonathan Cape: London.

Whiteley, S. (ed.) (1997) *Sexing the Groove: Popular Music and Gender*, Routledge: London.

Williams, R. (1982) [1958] *Culture and Society*, Hogarth Press: London.

—— (1993) [1962] 'Advertising: The magic system', in S. During (ed.) *The Cultural Studies Reader*, Routledge: London, pp. 320–36.

Williamson, Janice (1992) 'I-less and gaga in the West Edmonton Mall: Towards a pedestrian feminist reading', in D.H. Currie and V. Raoul (eds) *The Anatomy of Gender: Women's Struggle for the Body*, Carleton University Press: Ottawa, pp. 97–116.

Williamson, Judith (1978) *Decoding Advertisements: Ideology and Meaning in Advertising*, Marion Boyars: London.

Williamson, N. (1997) 'They came in search of paradise', *Observer*, 25 May, pp. 16–17.

Willis, P. (1977) *Learning to Labour: How Working Class Kids Get Working Class Jobs*, Columbia University Press: New York.

Wilson, E. (1995) 'The invisibile *Flâneur*', in S. Watson and K. Gibson (eds) *Postmodern Cities and Spaces*, Blackwell: Oxford, pp. 59–79.

Wright, M.A. (1999) Personal communication.

—— (1998) 'Ecstasy use and the British dance scene: The symbolic challenge for new cultural movements', unpublished Ph.D. thesis, Department of Sociology, City University, London EC1V 0HB.

—— (1997) 'Sociological effects', in N. Saunders (ed.) *Ecstasy Reconsidered*, with a bibliography by Alexander Shulgin, Neal's Yard Press: London, pp. 64–71.

—— (1995) 'Dance culture/music culture, in N. Saunders (ed.) *Ecstasy and the Dance Culture*, with contributions by Mary Anna Wright and an annotated bibliography by Alexander Shulgin, Neal's Yard Press: London, pp. 170–211.

Wrigley, N. and Lowe, M. (eds) (1996) *Retailing, Consumption and Capital: Towards the New Retail Geography*, Longman: Harlow, Essex.

Wyndham, H. and St. J. George, D. (1926) *Nights in London: Where Mayfair makes Merry*, The Bodley Head: London.

Yeats, W.B. (1956) *The Collected Poems of W.B. Yeats*, Macmillan: London.

Young, J. (1997) [1971] 'The subterranean world of play', reprinted in S. Thornton and K. Gelder (eds) *The Subcultures Reader*, Routledge: London, pp. 71–80.

Films

Coming Down (1997) Dir. Matt Winn.
Goodfellas (1990) Dir. Martin Scorsese.
Saturday Night Fever (1977) Dir. John Badham.

Music

D*NOTE (1997) *Coming Down*, Virgin Records (VCRD19).

The Gentle People (1997) *Soundtracks for Living*, Rephlex Records: London (rephlex cat045cd).

Massive Attack (1992) 'Unfinished sympathy' on *Blue Lines*, Wild Bunch Records: London (WBRCD 1).

Primal Scream (1991) 'Come together' on *Screamadelica*, Creation Records: London (CRECD 076).

Primal Scream (1991) 'Higher than the sun' on *Screamadelica*, Creation Records: London (CRECD 076).

Primal Scream (1991) 'Loaded', on *Screamadelica*, Creation Records: London (CRECD 076).

Shaw, Marlena (1977) 'Look at me, look at you' on *Sweet Beginnings*, Blue Note Records.

INDEX